YARN AND FABRIC FORMING
现代纺织英语

尚伟 编译

东华大学出版社·上海

图书在版编目(CIP)数据

现代纺织英语/尚伟编译. —上海：东华大学出版社，2019.1
ISBN 978-7-5669-1501-6

Ⅰ. ①现… Ⅱ. ①尚… Ⅲ. ①纺织工业—英语 Ⅳ. ①TS1

中国版本图书馆 CIP 数据核字(2018)第 262204 号

责任编辑：竺海娟
封面设计：魏依东

现代纺织英语

尚伟　编译

出　　　版：东华大学出版社(上海市延安西路1882号　邮政编码:200051)
本 社 网 址：http://dhupress.dhu.edu.cn
天猫旗舰店：http://dhdx.tmall.com
营 销 中 心：021-62193056　62373056　62379558
印　　　刷：常熟大宏印刷有限公司
开　　　本：710 mm×1000 mm　1/16
印　　　张：11.75
字　　　数：300 千字
版　　　次：2019 年 1 月第 1 版
印　　　次：2019 年 1 月第 1 次印刷
书　　　号：ISBN 978-7-5669-1501-6
定　　　价：35.00 元

前　言

本书较全面介绍了世界纺织工业在纱线与织物成形方面的近况及趋势，内容包括纺纱工程、化纤纺丝、织造工程、数据处理、空气工程、自动化、针织工程、电脑刺绣、纺织测试等方面。在内容上力求符合现代化的需要，并与目前的专业书刊相配合，以适应我国在国际上的技术交流、合资合作、国际展会日益频繁的现状。

本书是为有英语基础及纺织专业基础的读者编写的，内容以英语课文为主。对于比较难懂和一般英汉词典不易查到的词汇和短语，均有简要注释。为了便于读者阅读，本书分为阅读材料及参考译文部分。每篇课文后面附有练习，以促进读者对课文的理解。本书除供科技人员与管理人员学习纺织英语外，也适合作为纺织专业高年级学生的英语学习参考用书，以适应现代纺织技术对机电一体化复合人才的需要。

本书在编写过程中，主要参考了近几年《国外纺织技术》的内容及《国际纺织通报》中文版，以使参考译文更准确。沈阳市纺织科学技术研究所、沈阳丝织总厂等单位为本书提供了大量的技术资料与帮助，在此深表谢意。

由于编者水平与学识所限，书中错误和问题可能很多，恳请读者批评指正。

编　者

CONTENTS
目 录

Part 1 technical articles 课文

UNIT 1 Short staple spinning 短纤维纺纱
Lesson 1 Short staple spinning (1) ·················· 1
Lesson 2 Short staple spinning (2) ·················· 5
Lesson 3 Short staple spinning (3) ·················· 9

UNIT 2 Man–made fiber primary spinning 化学纤维纺丝
Lesson 4 Man–made fiber primary spinning (1) ·················· 13
Lesson 5 Man–made fiber primary spinning (2) ·················· 17

UNIT 3 Yarn treatment, make–up and finishing 纱线的加工、包装及处理
Lesson 6 Yarn treatment, make-up and finishing (1) ·················· 20
Lesson 7 Yarn treatment, make-up and fiqishing (2) ·················· 24

UNIT 4 Texturing and filament yarn processing 变形及长丝纱线的加工
Lesson 8 Texturing and filament yarn processing (1) ·················· 26
Lesson 9 Texturing and filament yarn processing (2) ·················· 29

UNIT 5 Machinery for wool spinning 毛纺机械
Lesson 10 Machinery for wool spinning (1) ·················· 31
Lesson 11 Machinery for wool spinning (2) ·················· 35

UNIT 6 Air engineering 空气工程
Lesson 12 Air engineering (1) ·················· 40
Lesson 13 Air engineering (2) ·················· 44

UNIT 7 Data processing 数据处理
Lesson 14 Data processing (1) ·················· 46
Lesson 15 Data processing (2) ·················· 49

UNIT 8 Automation of handling and transport 装卸与运输的自动化
Lesson 16 Automtation of handling and transport (1) ·················· 52
Lesson 17 Automation of handling and transport (2) ·················· 56

UNIT 9　Weaving preparation 织造准备
Lesson 18　Weaving preparation ·· 58
UNIT 10　Shed-forming motions 开口形成机构
Lesson 19　Shed-forming motions ·· 61
UNIT 11　Development trends in weaving 织造的发展趋势
Lesson 20　Development trends in weaving（1） ································ 65
Lesson 21　Development trends in weaving（2） ································ 69
Lesson 22　Development trends in weaving（3） ································ 73
Lesson 23　Development trends in weaving（4） ································ 77
UNIT 12　Machines for carpet manufacture 制造地毯的机器
Lesson 24　Machines for carpet manufacture ···································· 80
UNIT 13　Testing and measuring equipment 试验与测量仪器
Lesson 25　Testing and measuring equipment（1） ······························ 84
Lesson 26　Testing and measuring equipment（2） ······························ 88
UNIT 14　Flat knitting maching 针织横机
Lesson 27　Flat knitting maching（1） ·· 92
Lesson 28　Flat knitting maching（2） ·· 95
UNIT 15　Large-diameter circular knitting machines and ancillaries 大直径圆型针织机及辅助装置
Lesson 29　Large-diameter circular knitting machines and ancillaries（1） ·········· 98
Lesson 30　Large-diameter circular knitting machines and ancillaries（2） ········ 102
Lesson 31　Large-diameter circular knitting machines and ancillaries（3） ········ 106
UNIT 16　Warp-knitting and related machinery 经编及相关机械
Lesson 32　Warp-knitting and related machinery（1） ·························· 109
Lesson 33　Warp-knitting and related machinery（2） ·························· 113
UNIT 17　Embroidery 刺绣
Lesson 34　Embroidery（1） ·· 116
Lesson 35　Embroidery（2） ·· 120
UNIT 18　Braiding and bobbin lace machines 编带及梭结花边机
Lesson 36　Braiding and bobbin lace machines（1）flat knitting machined（1） ·· 124
Lesson 37-Braiding machines（2） ·· 127

Part 2　Translations 参考译文 ··· 131

Part 1 technical articles 课文

UNIT 1 SHORT STAPLE SPINNING 短纤维纺纱

LESSON 1
SRORT STAPLE SPINNING(1)

In recent years, for the short-staple spinning sector not ITMA① has not exhibited new processes and machines. it was rather a question of existing processes being modified in association with increased productivity and quality upgrades. Automation activities remained stagnant at the already high standard achieved in industry. Process data collection and manufacturing control have only succeeded in surviving② where they are economically viable vocabulary. On the other hand technological and technical ideas have enjoyed a resurgence where they can make a contribution to quality improvement, higher productivity, reduction in maintenance requirements, and improving resource utilization.

Some new processes from ITMA'91 have also become firmly established.

Blowroom

It is now possible for automatic obliquely angled bale opening to be preceded by virtually automatic bale laydown. Following the automatic remove of bale bands and manual removal of bale wrappers, the obliquely operating detacher is loaded with the aid of a self-loading truck③. High flexibility provides for several

---

① ITMA：*International Textilmachinausstellung*（德语）国际纺织机械展览会
② ... succeed in + ing　成功地(做成某事)
③ with the aid of a self-Loading truck　借助于自动加载小车

bale opener systems to be served[①]. The entire system is of modular design, enabling automation to be implemented stage by stage.

The intelligence level of bale opening systems has been improved by computer control. There are provisions for the automatic loading of batches from rectangular laydown to oblique laydown and concluding without any residual fiber, It is no longr necessary to level-off bales of different heights[②] before starting a blend. Secondly, the detacher requirement (production rate) is controlled automatically by the detacher increment. The speed controls of detacher systems are equipped with frequency-con-trolled motors, permitting quick adaptation to the fibers involved without having to change gearings.

The throughput of such bale opening systems today extends to 1600 kg/h. A number of blends can be serviced by one detacher system with swiveling turrets. Control and malfunction detection provisions are available on bale openers.

In cotton cleaning the concept of multi-roll cleaners arranged in tandem (3~4 rollers in succession) has become generally established. The cleaning principle of high centrifugal force of particles at relatively low roller peripheral speeds appears to represent the optimum in terms of cleaning efficiency and limitation of fiber damage.

Marzoli[③], like Trutzschler[④], has means for modifying the level of cleaning by stepless adjustment of the knives via servomotors. In addition to the knife clearers, carding plates are provided for opening purposes. Rieter[⑤] continues to favor non-damaging cleaning without direct fiber clamping and with computer control of cleaning intensity.

---

① High flexibility provides for sevveral bale opener system to be served 高度的灵活性为若干抓包机系统服务创造了条件。

② It is no longer necessary to level-off bale of different heights... ……不再需要调整不同高度的棉包的水平位置。

③ Marzoli （意大利）马佐里公司

④ Trutzschler （德国）特吕茨施勒尔公司

⑤ Rieter （瑞士）立达公司，本课图片为该公司 Varioset 清花间概念的一部分

All systems provide rapid adaptation of machine settings to the particular raw materials situation and the level of cleaning required. Strenuous development efforts have been made back in the blowroom with regard to the detection of extraneous fibers. Modern image processing facilities combined with powerful computer techniques permit high-speed analysis of successive images produced with incident, transmitted and reflected light. This permits detection not only of particles of different color but also those of different reflectance characteristics. A further ensuring minimum fiber loss is the provision of quick acting closure flaps with response time of 50 ms.

The idea of extraneous fiber detection in the blowroom appears to have proved an essential feature of the spinning plant. The short cotton cleaning line with a minimum number of beater points[①] has now become reality.

Vocabulary

viable 可行的，生存的
on the other hand 在另一方面
resurgence 复苏，复活，再生
blowroom 清花间
precede 居先，位于，在
virtually 实际上，事实上
removal 移去，除去，卸去
upgrade 提高等级，提高质量
stagnant 停滞的，滞的
technological 工艺的
utilization 利用，应用
obliquely 斜地，不正地；……之前
laydown 放下，搁置
wrapper 包布
flexibility 灵活性，适应性

detacher 抓棉器
truck 小车，抓棉小车
modular design 积木化设计
provision 设备，机构，装置
rectangular 矩形，成90°角
residual 残留
level-off 使成水平，调整水平位置
increment 增加，增量
gearing 齿轮传动，传动装置
swivel 旋转，摇摆
malfunction 故障，出错
available 现有的，存在的，备用的
arrange 安排，装配，安装
particle 粒子，颗粒，质点
peripheral 外围，外部

---

① The short cotton cleaning line with a minimum number of beater points... 具有最少打击点的短流程清棉生产线……

clothing 针布
servomotor 伺服电机
clamping 握持，夹紧
strenuous 艰苦的，努力的
facility 设备，装备
successive 连续的，顺序的
transmit 发射，辐射
permit 允许，使得有可能
adaptation to 适宜……，适合于……
in addition to 除……之外
bale opener 开包机，抓包机
provide 为……创造条件，提供，供给
implement 执行，实现，完成
batch 一批
conclude 结束，终止
no longer 不再
blend 混合，混纺纱
frequency-controlled motor 频控电机
throughput 原料通过量，生产量

turret 转塔，转动架
service 运行，操作
tandem 串联，串列，串列布置
centrifugal force 离心力
principle 原理
optimum 最好的，最宜的
knife 除尘刀片
carding 梳棉，梳理，粗梳
intensity 应力，强度
extranecus 外来的，外部的
image 成象，反射信号
incident 事件，差错
beater point 打击点
processing 处理，加工
in term of 就……而论
with regard to 关于，对于
combined with 与……结合，与……分不开

Think or answer these questions

1. How much do you know about ITMA?
2. By what has the intelligence level of bale opening systems been improved?
3. How much is the throughput of today's bale opening systems?
4. Can the carding plates be used for opening purposes?
5. Which techniques and facilities are needed to detect the extraneous fibers?

LESSON 2
SHORT STAPLE SPINNING(2)

Cards

The greatest technological advances in terms of quality and increased productivity are at present to be found in the carding sector. For instance the number of card flats present in the operating position is being reduced and replaced by stationary flats. The main function today of the revolving flat with flexible clothing is primarily to remove short fibers and residual trash, to open-up the fiber and also remove neps and trash.

Opening is undertake by upstream stationary carding segments or flats. These carding elements with metallic clothing can manage higher workloads than flexible flat clothing. Card makers have now developed customized elements for their cards. All carding elements now possess appreciably higher precision and greater adjustment facilities than they used to. These permit more discriminating reproducible setting in long-term operation. All these measures have enabled the number of revolving flats to be reduced to 21 in the carding position. Hollow flats profiles have greater precision than cast flats in terms of sag characteristics and carding performance is constant across the full machine width.

Trutzschler's arrangement of three takes-in① sees the revival of an old idea. The pressure for ever greater productivity calls for measures for reducing the load on carding elements in the cylinder-and-flats zone. The fibers leave the take-in zone with fiber tufts very substantially opened up into web form. With such devices card production rates in excess of 140kg/h are realistic with good carding quality. In turn, with a given raw material and known production rate, carding quality is improved resulting in better utilization of resources. The useful life of card clothing is also being extended.

The arrangement of three takes-in is a logical step towards long-term higher productivity on the card. The card is thus increasingly incorporating the functions of the blowroom, since with a given throughput the three-part arrangement ②is eminently able par-

---

① three takes-in　指特吕茨施勒尔公司 DK 903 梳理机的"Webfeed"三刺辊系统，它代表新型梳理机最值得注意的部分，最新型号的 DK 903 梳理机仍旧采用三刺辊系统
② the three-part arrangement　三刺辊排列

tially to undertake the cleaning function via adjustable knives.

But the high-tech① card now more than ever calls for controlled reproducible adjustment. For the cylinder-and-flats zone there are now facilities for precision flats setting at different setting sites using the Trutzschler Flat-Control② or the Card-Setting-System③ developed by ITV Denkendof④. The setting is displayed by electronic sensors. Reporting facilities ensure that the card can be reset with its initial settings following repeated servicing or grinding. Settings are thus free from subjective influences. The ITV system can be adapted to any card. Concentricity and expansion of the cylinder can also be determined. Trutzschler has also developed on-line nep counting⑤ to full production standards. Objective monitoring of carding quality is now a possibility.

The tandem card by Crosrol⑥ continues to represent an extension of the three-part take-in arrangement. Each flat is individually adjustable in height. This reduces or eliminates the need for grinding the set of flats. The hardened points are retained which extends the useful life.

Modern card design thus permits better resource utilization by upgrading yarn quality and increasing productivity. Better resource utilization provides the most economically

---

① high-tech　tech(technology 缩写)高技术
② Flat-Control　盖板隔距测量系统
③ Card-Setting-System　ITV 的梳理机调节系统
④ ITV Denkendof　(德国)邓肯道夫纺织工艺技术研究所,简称 ITV
⑤ on-line nep counting　在线棉结计数
⑥ Crosrol　(英国)克诺思诺公司

viable basis for capital investment in the carding sector.

Hollingsworth① showed a system of carding using solely stationary carding segments, the new Clean-Master-System② comprising six narrow carder elements with knives is extremely flexible in terms of clothing and settings. Practical trials confirm this system can now also be used successfully in cotton processing. The future will show whether not only the partial but complete elimination of revolving flats with flexible clothing will be a practical possibility.

Drawframes

The simple three-over-three③ drafting system is probably the most widely built system and thus the most successful drafting arrangement for short-staple processing. The new Trutzschler drawframe employs a multi-motor drive for entry and exit including sliver evening. Mechanical sensing is retained but with a grooved condenser. Present-day evener systems and quick-response servomotors permit correction lengths of about 10 cm in the event of sliver breakage. The rectangular can is emerging as a real alternative to the circular can④. Emanating from rotor spinning, it now comes into consideration for speed-frame creeling, and as feed for the air-jet spinning machine. The superior space utilization and easier handling in automation situations favor the rectangular can in the final drawing stage. Drawing stages can be joined up fully automatically by linking system and drawing speeds of 1100m/min are reality.

Comber

An entirely new comber incorporating technologically important new concepts was offered by CSM⑤. Central reproducible feed setting for all eight comber heads without change gears increase ease of operation. A flexible cushion plate with pneumatic weighting adapts automatically to irregularities in the lap and sliver formation within the lap. Associated with this is improved fiber control at the moment of detaching which permits feed lap weights of up to 100 ktex. The cooperating detacher roll and pneumatic support

---

① Hollingsworth （美国）豪林沃思公司
② Clean-Master-System 豪林沃思公司的梳理系统名称
③ three-over-three 三上三下
④ (be)an alternative to M 是 M 的替代物，可以替代 M
⑤ CSM Chmnitzer Spinnereimashinenbau 的缩写，CSM 公司

of the combed fringe improve sliver piecing. Nip rates of up to 500/min appear to be quite practicable. Automatic feed from sliver lapper to comber is possible by robotic conveyors. Nip rates of 400/min are generally practicable today. Rieter still has the only comber with fully automatic lap replenishment.

Vocabulary

for instance 例如
card flat 梳理机盖板
revolving flat 活动盖板
flexible clothing 弹性针布
residual trash 残留杂质
undertake 承担，单任
segment 部分
workload 工作负荷，工作量
appreciable 明显的，可观的，相当的
reproducible 重复的，能再现的，可再生的
measure 方法，措施，手段
hollow 空心的
sag 下垂
constant 恒定，均匀，常数
revival 复活，复兴，再生
tuft 碎毛块，棉茸
in excess of 超过，多于
utilization 利用，应用
incorporate 结合，包括，组合
partially 部分地，局部地
grinding 磨的，磨削，研磨的
subjective 主观
expansion 匀整
probably 大概，或许

sliver 条子，纱条，梳条
condenser 集合器，集棉器
breakage 断裂，破损，断头率
rotor spinning 转子纺纱，气流纺纱
speed frame 粗纱机
superior 高级的，优秀的
join up 联合起来
stationary flat 固定盖板
nep 棉结
open-up 开放
upstream 上游，逆流
element 元件，单元，零件
customized 定做，定制
precision 精密度，精确性
discriminating 形成区别，有差别的
profile 轮廓，外形，剖析
cast 铸造，铸件
performance 性能，实行，操作
take-in 梳理机刺辊
arrangement 排列，布局，安装
substantially 本质上，坚强地
in turn 依此，而
logical 合理的，逻辑的
eminently 突出地，著名地

display 显示，指示，表示
initial 初始的，开始的
concentricity 同心度
tandem card 双联梳理机
employ 使用，应用
groove 沟槽
evener 条子自调均匀装置，均棉器
rectangular can 矩形条筒
emanate from 发源，发出
creeling 换条筒，换筒子，换粗纱
handling 加工，搬运，装卸
entirely 完全地，彻底地
central 中央，集中，中心

irregularity 不匀率
improve 改进，改变，改善
detacher roll 分离罗拉
lapper 条卷机
piecing 接头
replenishment 再补给，再补充，供给
cushion plate 梳棉机下钳板
lap 毛卷，棉卷
at the moment 现在，那时
cooperate 合作，配合
fringe 网边
nip rate 钳次速度

Think or answer these questions

1. What is the main function today of the revolving flat with flexible clothing?
2. What advantages has the arrangement of three takes-in?
3. Will elimination of revolving flats with flexible clothing be possible?
4. What advantages has the rectangular can?
5. Can the cushion plates be with pneumatic weighting?

LESSON 3
SHORT STAPLE SPINNING(3)

Speedframes

  The most advanced speedfram design displayed was the Grosenhainer[①] BF224 cotton speedframe. The multi-motor drive continues to replace all mechanical drives apart from draft changing. Tne change in flayer pitch to 224 mm eliminates the familiar pitch

---

① Grosenhainer （德国）高诺森海纳公司

gap on the machine. Some speedframes have sliver sensor control.

The Zinser speedframe is another operating without change gears and with mechanical/pneumatic control elements. Tried-and-tested① mechanical modules such as cone gears and package build have high-precision control. All speedframe have automation options for both doffing and package replenishment.

Ring spinning

Of all the stages of spinning, ring spinning was the one in which consolidation of the rapid developments of recent years was most marked. Super-long machines with up to 1632 spindles involved new concepts in spindle drives. Along with the successful four-spindle drive systems sectional drive with main drive shaft and narrow drive belts appeared. Sectional belts drive 48 spindles in the case of the Suessen system and are endless. Replacement presents no problem. A new drive concept was proposed by Zinser. Even on long machines the frame has only one tangential belt which is driven in sections by up to seven motor. The belt thus has only one speed and can be kept very narrow, resulting in an extremely low noise level. The drafting mechanism bearings can be adjusted precisely and steplessly along the entire length, providing maximum precision in roller mounting on long machine.

The Cerasiv② ceramic spinning ring developed by ITV Denkendorf represents an innovation in ring spinning. This spinning ring eliminates costly ring running-in programmes at present-day traveler speed. Secondly, the purposely developed travelers give longer useful life and reduce the frequency of the traveler changing. Industrial trials have shown traveler life to be 2.5 to 3 times longer than in today's customary systems. The system manages without lubrication in worsted spinning. There are serious advantages here in respect of yarn quality, lint, servicing and durability. The operation of the ceramic spinning ring was demonstrated at ITMA on the CSM worsted ring spinning frame. In all sectors of ring spinning that are critical today, where travelerwear is an obstacle on the route to higher productivity, this system represents an initial stage in increasing productivity.

---

① tried-and-tested 经试验验证的
② Cerasive 在 Plochingen 的德国 Cerasive 公司，归属工业陶瓷领域。

Spinning without conventional triangular spinning zone ①(compact yarn spinning②) permits better fiber utilization through higher yarn strengths. The twist level can be also reduced. A feature of this yarn is its low hairiness and no lubricant is required for the ring/traveler system. In early experience with the compact yarn system, as already demonstrated by CSM in the case of worsted spinning, ceramic ring based ring and traveler systems are playing an important role.

Unconventional spinning systems

In rotor spinning the familiar spin-boxes based on Elitex developments have been offered in recent years. Rotor speeds with direct rotor bearings extended to 90000 mit^{-1}. High-speed rotor spinning machines continue to be the machines with indirect bearings by Schlafhorst③ and Rieter. A speed approaching 1600 mit^{-1} is attained by Schlafhorst today, corresponding to an increase of around 28% compared with IMA'91. The performance limitations of the rotor spinning machine in the medium count range are due more to the take-up speed. The R1 machine made by Rieter with a yarn piecing system having fiber flow reversal permits joins to be produced that demonstrably increase productivity in weaving. A completely new rotor-box with maximum flexibility, the SE 10 Box④, was

---

① triangular spinning zone   纺纱三角区,它是除钢领、钢丝圈系统外在环锭纺纱中最薄弱的环节。
② compact yarn spinnineg   紧密纺纱,是最近几年发展起来的纺纱新技术,属现代环锭纺纱最先进的技术,具有纺纱技术前沿性。
③ Schlafhorst   (德国)赐来福公司(属Suarer集团),本课图片为该公司的Autoconer 转杯纺纱机。
④ SE 10 BOX, SE 9 BOX   绪森公司设计生产的纺纱器(供赐来福公司生产转杯纺纱机),最新型号 SE 12 BOX 由赐来福公司设计,具备纺纱科技前沿性。

YARN AND FABRIC FORMING

developed by Suessen. All spinning elements can easily be interchanged without any special tools. Adapting the feed rate to the rotor diameter is performed without removing the front panel. Energy consumption has been reduced and the familiar automation of the SE9, SE10, SE11, SE12 Box has been adopted with starter bobbins continuing to be used.

At 400m/min air-jet spinning attains the highest spinning speed of all staple fiber spinning systems. A new introduction was a machine with a single-jet followed by crossed rubber rollers for producing false-twist. These rollers produce weaving yarns of very low hairiness. Mixture yarns with a polyester content of at least 40% are also spun with 804 RJS①. The machines can be fitted with piecing devices as an alternative to knotters. In this case the yarn end from the yarn package is placed on the raised front roller and pieced-up at full speed. Air-jet spinning too is thus in constant pursuit of higher productivity. It remains to be seen② whether it will ever attain the desired level of flexiblitity.

Vocabulary

flyer 锭翼
package build 卷装成形
consolidation 强化，凝固，合并
sectional 分段
cone gear 圆锥轮
ring spinning 环锭纺纱
marked 明显的，显著的
up to 直到，等于
tangential belt 龙带，正切传动带
ring 环，钢领
purposely 故意，特意地
lint 飞花，棉绒

wear 磨损
route 道路，路线，方法
due to 由于
fiber flow 纤维流
adopt 采用
automation 自动装置
mixture yarn 混纺纱
at full speed 以全速
(be) compared with 与……比较
(be) followed by 在……加上
mechanism 机构，装置
running-in 试车，试运转，校车

---

① 804 RJS　RJS(Roller Jet Spinning 的缩写)日本村田公司生产的一种罗拉喷气纺纱机。
② It remains to be see... ……尚待观察。

serious 重大的，严重的
service 维护，照料
obstacle 障碍，妨害
medium count 中支[数]
take-up 卷取
reversal 反向，倒转
front panel 前部面板

starter bobbin 打低筒子
introduction 引进，推广，采用
knotter 打结器
correspond to 和……符合，相当于
adapt to 使……适应
in pursuit of 为了求得

Think or answer these questions

1. What changes have taken place in the most advanced speedframe design?
2. What advantages has the ceramic spinning ring?
3. What are the performance limitations of the rotor-spinning machine in the medium count range due more to?
4. Which one attains the highest spinning speed of all staple fiber spinning systems?
5. Are mixture yarns also spun with air-iet spinning machines?

UNIT 2 MAN-MADE FIBER PRIMARY SPINNING
化学纤维纺丝

LESSON 4
MAN-MADE FIBER PRIMARY SPINNING(1)

Filament yarn

There have been very many suppliers of synthetic filament yarn machines in recent years. With the continuing improvements made to machines, which at one time were only suitable for polypropylene, most engineering companies offer machines which those will spin any of the major polymers.

The two main themes now seem to be energy saving and minimum process interrup-

YARN AND FABRIC FORMING

tions. This has led most companies to opt for bottom loading packs[1] which avoids the heat loss associated with the chimney effect caused by top loading. Process interruptions are reduced by, for example, better polymer filtration through continues filters and by fitting turret winders which doff without an interruption or any yarn waste.

Although winders are available for making undrawn yarn, now known as low orientation yarn or LOY[2], more emphasis is being placed on lines which produce POY[3], HOY[4](high oriented) and FDY[5](fully draw yarn).

– **Interlacers**

Many companies stress the high efficiency of the interlacers fitted to their lines. This reduces interruption, gives better package builds and, most importantly, reduces the amount of size required for subsequent operations. This benefits the environment.

– **Winders**

The development of winders seems to be have been directed at longer chucks which will wind more, or wider, package. Typical was Barmag's[6] CW b 1500 winder which

---

① bottom loading packs 下装式喷丝头组件
② LOY 低取向丝
③ POY 预取向丝(高速纺丝)
④ HOY 高取向丝(超高速纺)
⑤ FDY （Full Draw Yarn）全拉伸丝
⑥ Barmag（德国）巴马格公司

had a 1500 mm chuck length and would take 6, 8 or 10 tubes. The winder produces POY at speeds between 1500 and 4000m/min.

– **POY**

As a result of the acquisition of Automatik GmbH① in 1992, Rieter was able to offer, for the first time at an ITMA, melt spinning machinery②.

Of special interest was a POY machine with novel features such as spinning pumps mounted above the spinning beam③ and a Dowtherm boiler mounted next to the beam. The latter, together with other heat saving design steps, was said to cut heat losses by 25%.

– **FDY Yarn**

One very valuable development for the production of full-drawn-yarns(FDY) was to be made by John Brown Deutsche Engineering④. The convetional FDY machine uses godets to stretch the yarn whereas the John Brown hot channel stretching machine⑤ introduces orientation into polyester filaments by heating the filaments in a counter current of hot air. Lower costs, better uniformity, less breaks and uniform dye uptake a claimed for⑥ this process. The winder speed used is in the range 5000~6500m/min.

The other unconventional FDY process is the EMS-Inventa's⑦ H4S⑧ process. In this process unheated godets stretch the yarn and a steam chamber simultaneously interlaces and relaxes the filaments.

– **Industrial Yarns**

Industrial yarns from polyesters and polyamide polymers are being produced on one-step machines. For normal polyester yarns Barmag recommend speeds of up to 4000m/min whereas for high-modulus-low-shrink (HMLS) polyester yarns, bow being used in tyres, speed of up to 8000m/min are used.

---

① Automatik GmbH Rieter 集团的 Automatik 公司
② melt spinning machinery 熔体纺丝机械
③ spinning beam 纺丝箱体
④ John Brown Deutsche Engineeri 约翰. 布朗德国工程技术公司
⑤ hot channel stretching machine 热管拉伸机
⑥ a claim for 宣称……为其所有
⑦ EMS-Iventa （瑞士）依姆斯-因维塔公司
⑧ H4S 冷辊拉伸蒸汽定形法生产工艺

YARN AND FABRIC FORMING

现代纺织英语

On the other hand for the highest strength polypropylene industrial yarns, the UK-based Extrusion Systems Ltd①. and PFE Ltd., both recommend the use of two stage processing. Tenacities of up to 10g/den are claimed which makes polypropylene a serious competitor to both polyester and polyamide in many applications.

Vocabulary

supplier 供应商，供应者
engneering 工程技术，技术设备
polypropylene 聚丙烯
polymer 聚合物
major 主要的，多数的
loading 装料，加料，加载
filtration 过滤
fitting 符合，装配，配件
known as 通称为，皆知为……
emphasis 重点，重要性，强调
stress 强调
subsequent 后来的，连续的
benefits 对……有利
chuck 筒管夹头
the latter 后者
stretch 拉伸
counter current 对流，逆流

polyamide 聚酰胺
high-modulus-low-shrink 高模数低收缩
competitor 竞争者
chimney 烟囱，尘塔
pack 组装，组件
filter 过滤器
undrawn yarn 未拉伸丝
orientation 取向，排列方向，定向度
interlacer 交缠喷嘴
efficiency 效率，功效
amount of 数值，大小，数量
efficient 有效的，效率高的
acquisition 获得，发现，探测
godet 导丝辊
dye uptake 上染率，上色率
recommend 推荐，建议
tenacity 断裂强度

Think or answer these questions

1. Which lines that produce filaments is more emphasis being placed on?
2. What advantages have the interlacers?

---

① UK-based Extrusio Systems Ltd.   UK-based Extrusio Systems 有限公司

3. Why dos the development of winders seem to have been directed at longer chuck?
4. What advantages has the john Brown hot channel stretching machine compared with the conventional FDY machine?
5. What industrial yarns are being produced on one-step machine?

LESSON 5
MAN-MADE FIBERS PRIMARY SPINNING(2)

– BCF

Three-end carpet machine were to be seen in a lot of places. Barmag claimed that its 3-end machine achieved an unprecedented cost/performance ratio①. Plantex② offered a new high speed BCF③ compact producing tricolour yarns from 1000 to 3600 denier at speeds up to 3000m/min. The machine could handle all types of melt spinnable polymers.

– Nano filaments

It is reported that Toray has developed world's first Nano Nylon Filaments④. The technique is suited for common polymer (such as polyamide, polyester), and can be produced on conventional machines.

– Staple Fiber

Staple fiber lines fall into two categories, two-stage and compact one-step⑤. The newer compact machines⑥ cover all process steps, from supplying granules to baling in one continuous operation. Fleissner⑦ says that its compact machine is suitable for

---

① cost/performance ratio 成本/性能比率
② Plantex （意大利）普兰泰克斯公司
③ BCF 把纺丝、拉伸、变形工序都在一台纺丝机上完成（简称 BCF）
④ Nano Nylon Filaments 纳米尼龙纤维，纳米技术为当代科技前沿性技术
⑤ two-stage and compact one-stage 两步型与紧凑一步型
⑥ compact machine 紧凑型机器
⑦ Fleisser （德国）弗莱斯纳公司

spundyed fibers because color shade can be checked quickly and both Fleissner and Neumag① claim that even small lots of staple fiber can be produced economically on such machines. Fleissner, in addition to supplying lines for polyester, polyamide and polypropylene, is a company which also offers wet and dry spinning polyacrylic lines.

Today the number of holes in each spinneret for staple fiber spinning is typical several thousand such us EMS-Inventa's micro-staple line which has 6000. As spinnerets get bigger and on the Zimmer stand② one saw a semi-automatic handling unit for the self-centering self-sealing pack said to③ weigh 250 kg. This company has also developed a new "jumbo" crimper which will enable a single 200t/day production line to operate with one crimper only.

What used to be called "novel" or "*speciality*" fibers④ are no longer a novelty or a speciality. Several companies now offer spin packs/spinnerets to make such fibers. Neumag, for example, offer equipment to make side/side⑤ and core/sheath bicomponents, hollow fibers and even hollow bicomponent fibers.

Particularly impressive bicomponent technology⑥ was developed by b. g. plast⑦. A compact machine was spinning polyolefin bicomponent staple fiber through a spinneret containing 20000 holes and 30000 holes are said to be possible.

Toray⑧ offers the production technique of sea-island composite superfine staple⑨.

– Conclusions

The fact that so many engineering companies are offering such a wide range of good quality machinery for producing man-made fiber will make decision making very difficult for potential customers. The brochures of marry companies therefore emphasise the long experience the company has had in the field and this must be an important consideration

---

① Neumag （德国）纽马克公司
② Zimmer stand （德国）吉玛公司的展台
③ be said to + inf 据说，被说成，被认为
④ *speciality fiber* 差别化纤维（来源于日本的外来语）
⑤ side/side 并列型
⑥ bicomponent technology 双组分技术
⑦ b. g. plast b. g. plast 公司
⑧ Toray （日本）东丽公司
⑨ sea-island composite superfine staple 海岛型复合超细短纤维

for would be buyers.

On the other hand one can but feel that delivery time, technical assistance, guarantees and above all price will be major factors.

Vocabulary

three-end carpet 三股纱地毯
plant 设备,装置,工厂
handle 处理,操作
category 种类,类型
baling 打包
shade 色泽,色光,明暗程度
crimper 卷曲箱
impressive 给人印象深刻的
brochure 小册子
can but + inf 只能
unprecedented 先例的,空前的,从未有过的

compact 小型的,紧凑的
melt spinnable polymer 可熔纺聚合物
granule 小颗粒
spun-dyed 纺前染色的,原液染色的
check 检验,控制,校验
micro-staple 微短纤维
pack 纺丝头组合件,组件,组装
particularly 特别,尤其,详细地
conclusion 结论

Think or answer these questions

1. What is the denier range of the tricolor yarns produced by a BCF compact plant offered by Plantex?
2. Which two categories do staple fiber lines fall into?
3. Why is the compact machine very suitable for spun-dyed fibers?
4. Why is what used to be called "novel" or "speciality" fibers no longer a novelty or a speciality?
5. What is the reason that makes decision-making very difficult for one to by a set of machinery for producing man-made fibers?

UNIT 3 YARN TREATMENT, MAKE-UP AND FINISHING 纱线的加工、包装及处理

LESSON 6
YARN TREATMENT, MAKE-UP AND FINISHING(1)

Before dealing with new products in the twisting sector and discussing just a few of the trends observed at ITMA, we will briefly indicate the trends evident in all branches of textile engineering, namely improvements of detail, widening use of microprocessors, individual drives for winder heads and spindles, increased data recording and monitoring, operation manuals on CD-ROM① one purpose of which is to speed-up the identification of machine malfunctions and trouble-shooting, improvements in automated handling, and no process automation "at all cost"②.

Assembly winders

The trend in assembly winders towards precision-wind continued. This produces higher packing densities and higher package weights. In yarn feed to two-for-one twister③ there is still competition between single-thread and assembly wound packages. The arguments for and against the two systems remain the same. Mill practices must be determined after careful analysis of the advantages and drawbacks of the two systems.

Core wrapping and effect yarns

In core wrapping and effect yarns there are no fundamental innovations apart from improvements of detail. The progressive introduction of microprocessors for control of

---

① CD-ROM 多媒体计算机的只读光盘存储器
② no process automation "at all cost" "不惜一切代价"实现工艺自动化
③ two-for-one twister 倍捻机

machines and their motions permits virtually unlimited design opportunities.

Winding machines for point-of-sale make-up

The main feature of winders for point-of-sale make-up is their high standard of automation, which has progressed still further in recent years. Precision length measurement, ease of operation with electronic control systems and facilities for additive application are taken for granted. Point-of-sale make-up involves many different formats and types of cones and cylindrical cross-wound package, plus ball-winders and machines for producing hanks and skeins. A new machine offered by SSM[①] was the "preciflex" cross-winder for winding filament yarns.

With the control unit it is possible to programme centrally via a touch-screen virtually all parameters of a yarn package. This high standard of automation enable all conceivable types and formats of packages to be produced easily, not only in point-of-sale form but also for example for dye packages.

Continuous treatment

In continuous treatment of running yarns such as setting and bulking there have not been any significant changes although certain variants in individual machines were to be seen.

Twisting systems

Today's twisting systems include ring twisting, two-for-one twisting and balloonless twisting which used to be called two-stage twisting. Ring twisting is still of significance only in a few specialist sectors such as ths effect twisting mentioned earlier. Two-for-one twisting which has grown in popularity over recent years now has a noteworthy competitor in the new machine now described.

A real innovation seen for the first time at ITMA' 95 was exhibited by Hamel(Saurer Group)[②]. This is the "tritec-Twister"[③] first launched in 1993. in principle it is a balloonless twister derived from the familiar Hamel 2000 two-stage twister and uptwister with two nesting counter-rotating pots. Three times the rotational speed of the spindle

---

① SSM （瑞士）SCHARER SCHWEITER METTLER AG 公司简称
② Hamel(Saurer Group)　哈梅尔公司(属绍雷尔集团)
③ Tritec-Twister　（瑞士)哈梅尔公司研制的三倍捻机，加捻锭子转一转，可在纱线上加三个捻回，适用于任何种类的短纤纱加捻，具备纺织科技前沿性。

actually rotating is achieved, in other words one rotation of the spindle produces three turns in the final yarn.

As in the predecessor machines the threads are subjected to little stress since thread friction in the pots is very low. Along with the higher productivity, a further advantage claimed by the maker is low energy consumption. Automatic air threading is incorporated in the machine, activated by a foot pedal.

Vocabulary

make-up 包装
evident 明显的，显然的
trend 趋势，动向，发展方向
namely 即，就是，换句话说
detail 细节，部分
speed up 加快速度
packing density 排列密度
assembly wound（两步法捻线机的）再捻络筒

determine 确定，决定
core wrapping 包芯纱
motion 运行，运转
opportunity 机会
in some time 有时侯
additive 附加的，辅助的
ball-winder 绕球机
skein 绞纱，绞丝
centrally 在中心，集中

programme 编程序，程序
type 型号，种类，类别
form 形式，形状，外形
setting 定性
variant 不同的，变异的，变形
balloonless twisting 无气圈加捻
noteworthy 显著的，值得注意的
uptwister 上行式捻线机
turn 捻回，转，圈
stress 应力
threading 穿纱，引线
activate 驱动，起动，开动
be subjectde to 受到，经(承)受
argument for (against) 为赞成(反对)
……而辩论 manual 手册，说明书，指南
identification 识别，辨别
two-for-one twister 倍捻机
practice 实际应用，实施
drawback 缺点，缺陷
progressive 逐步的，逐渐的
virtually 事实上，实际上，实质上

facility 设备，(附属)装置
cross wound package 交叉卷绕
hank 绞
plus 加，外加，加上
touch-screen 触屏，触摸显示屏
conceivable 可以想象的，可能的
format 规格，大小尺寸
running 流动的，运行着的，运转着的
bulking 膨化变形
ring twisting 环锭捻线，下行式捻线
still 静，不动，无声
grow 渐渐变成，渐渐……起来
launch 提出，开始
nesting 嵌套，套装，套用
predecessor 以前有过的东西，原有
(事)物
incorporate 包括，安装有
foot pedal 脚踏板
deal with 论述，处理
be derived from 由……而来，源出于

Think or answer these questions

1. What direction does the trend in assembly winders towards?
2. What is the main feature of winder for point-of-sale make-up?
3. Which two kinds of machines does the "Tritec-Twister" derive from?
4. Why are the threads subjected to little stress on the "Tritec-Twister"?
5. What systems do today's twisting systems include?

LESSON 7
YARN TREATMENT, MAKE-UP AND FINISHING(2)

Two-for-one twister

In two-for-one twisters there is a tendency for two types of machines to be offered according to market demand. For high-volume production there is a simpler narrower machine, more specific to the yarn being twisted. The other type of machine with more sophisticated technology is intended for more versatile use and for specialty products.

The systems linking automatic winder or assembly winder with two-for-one twister exhibited at ITMA'91 in the two-for-one twisting sector were no longer on show in recent years. At the time they involved automatically-operating bobbin changers both for replenishing the two-for-one twister with its complex threading process and for doffing the fully wound packages. Even at that time it was not possible to claim positive economic viability for a linking system such as this.

The industy has not accepted this form of automation. On the other hand, automated handling of supply packages, twister packages and empty tubes has been perfected and threading tools are available.

Steaming and humidification

The steaming and humidification of yarn packages is in itself far from being a new technology and apart from a few improvements of detail has been unchanged in recent years. Both Welker[1] and Xorella[2] in this sector, presented results of research confirming the benefits of increasing the moisture content of cotton yarns. For insance Welkrer in a contract trial involving processing through to the weaving machine was able to prove that yarn performance can be significantly improved. Xorella commissioned research on the physical properties of cotton yarns and reaches the conclusion from the superior yarn

---

[1] Welker   Welker 公司
[2] Xorella   (瑞士)Xorella 公司

strength and elongation characteristics that there are benefits in downstream processing. Yarn steaming and humidification thus has greater signifincance than simply adding weight to the yarn. The economic viability of such systems needs to be considered primarily from the aspect of quality enhancement and consequently improved yarn performance and secondly increased weight. In the economic analysis it is of course necessary to include the energy consumption in steaming, which can vary according to the system used.

To sum up in this particular sector, it may be said that with current trends and with the engineering developments in recent years mills are better equipped to accomplish their task.

Vocabulary

high-volume 高容量，高容积
involve 涉及，含有，包括
positive 确定的，肯定的，完全的
versatile 通用的，多用途的，多功能的
changer 变换器
viability 生存性，生存能力
steaming 汽蒸
in itself 在本质上
moisture content 含湿量，含水量，含水率
through to 直到
enhancement 提高，增加
downstream 顺流
humidification 给湿，增湿作用
far from 远非，并不是
contract 收缩，缩水
commission 委托

Think or answer these questions

1. What two types of machines are needed according to market demand in two-for-one twisters?
2. By what factor is the automation limited in the two-for-one twisting sector?
3. Is the steaming and humidification of yarn packages in itself a new technology?
4. Can the benefits of increasing the moisture content of cotton yarns be confirmed?
5. How to do the economic analysis about steaming systems?

UNIT 4 TEXTURING AND FILAMENT YARN PROCESSING
变形及长丝纱线的加工

LESSON 8
TEXTURING AND FILAMENT YARN PROCESSING(1)

Draw-texturing machines

One consequence of the development of draw-texturing machines with increasingly high production speeds is than heaters and quenching units have become longer. this has resulted in high machines with restricted accessibility. This trend has been arrested in recent years with the introduction of the short high-temperature heater (HT-heater). All machine makers offer automatic bobbin doffing at every take-up winder head. As may be expected, these developments have led to new machine configurations.

– **HT-heater in the primary zone**[①]

With the introduction of the HT-heater in the primary zone the heating zone is reduced from approx. 2.5m to less than 1m. Temperatures in the HT-heater are between 300℃ and 500℃ with setting temperature heing about 200℃. Heater length, heater temperature and thread velocity must be inter-balanced in such a way that the setting temperature of the yarn is reached precisely at the heater exit. With its shorter heater

---

① primary zone　主伸区

the new machine configuration permits the structure to be lower in height with easier accessibility. Additionally, energy economies are achieved by the improved insulation and reduction of radiating surfaces.

– Integral automated bobbin doffing

Up to the present time, the take-up winder units have been located in the center of the machine, i. e. immediately in front of the secondary heater. In this way it is possible to service a machine side from an aisle. Several machine makers are now offering machines with modified configuration. The take-up units incorporating automated doffing are located on the opposite side to the service aisle. This means that there is a service aisle plus a second aisle for bobbin removal and tube replenishment for each machine side. Since machines of this configuration occupy more space, some makers have retained the former configuration with one aisle per machine side. It may be assumed that integral automatic doffing and consequently also two aisles per machine side will become the accepted practice. The approach to filament threading varies quite considerably. Solutions are both simple and elaborate. A development opportunity still exists here for automation.

Single-heater machines for producing high-elasticity yarns and air-bulking machines[1] are, for reasons of cost, derived from components of corresponding two-heater machines.

– Friction-twisting units

In friction-twisting units, disks 9 mm thick have become standard practice. The choice of disk material depends on the intended endues for the textured yarns. The problem of individual drive to the twisting element has been resolved. This advance in development has considerably reduced noise levels (approx. 12 dB A). It remains to be seen whether the economics of this system will be found acceptable. Temco[2] offered a unit that can be switched from Z to S twist. Other makers developed interchangeable modules for triaxial twister elements. These developments all contribute towards making the texturing process more adaptable. Muratec[3] employs a unit which simultaneously produces

---

[1] air-bulking machine 空气膨化机
[2] Temco Temco 公司
[3] Muratec （日本）村田公司

adjacent S-twist and Z-twist threads. The two yarns of different twist direction are intermingled and taken-up. This creates a textured filament yarn with no snarling tendency. Following the texturing process it is possible to add an intermingling jet in order to improve the withdrawal characteristics and possibly eliminate① the need for sizing.

All machine makers are using online measurement②, generally thread tension metering heads, for monitoring the texturing process.

Vocabulary

draw-texturing machine 拉伸变形机
quenching 骤冷，淬火，熄灭
heater 加热器
accessibility 可达性，易接近性
velocity 速度
integral 整体的，合成的
assume 假定，设想
endues 最终用途，产品用途
triaxial 三维的，三轴的，三向的
intermingle (互相)混合
withdrawal 退绕
i.e. *id est* (拉丁文)那就是，即

consequence 后果，影响，结论
increasingly 愈加，日益，格外
restrict 限制，限定
insulation 绝缘材料，保温，隔垫
aisle 通道，过道
corresponding 相当的，同位的，合适的
simultaneously 同时的，一齐地
adjacent 邻近的，相邻的，靠近的
snarling 缠结，纱线扭结
in such a way that... 通过……(方式)来

Think or answer these questions

1. How does the length of the heater have influence on the profile of draw-texturing machines?
2. Between what two points are temperatures in the HT-heater?
3. Do you think that the take-up winder units should be located in what place of the ma-

---

① eliminate the need for sizing 排除上浆的需要。长丝免浆技术作为一种高效、低成本的环保清洁加工技术，正在受到世界纺织工业的重视。
② on-line measurement 在线测量

chine?
4. How much is the thickness of the disks in today's friction-twisting units?
5. Which kind of measurement are all machine makers using for monitoring the texturing process?

LESSON 9
TEXTURING AND FILAMENT YARN PROCESSING(2)

Trends

The following general trends are discernible in texturing machinery:
-structures of lower height and easier accessibility
-energy economy and lower noise levels,
-more flexibility,
-on-line quality monitoring① and
-incorporation of the texturing machine in an overall automation programme.

Full oriented non-textured yarns② can on the one hand be produced by spinning on the FOY③ or FDY processes and on the other hand from POY using a supplementary draw process (draw-wind, draw-twist or draw-warp④)

Draw-wind and draw-twist becoming more important

The number of firms offering draw twist and draw-wind machines has increased further in 1995 compared with 1991. These systems are suitable for small batches and for speciality yarns. Draw-twist and draw-wind machines have up to now mainly been bought by manmade fiber producers and it is recommended that these machines should be used to a greater extent by textile processors too in the future. this will enable the textile processor to adapt the draw process to his own products in the best and most flexible way. POY yarns are available at low cost throughout the world.

Zinser offers "Co-We-Mat⑤" automatic doffing for the draw-twist machine and is a-

---

① on-line quality monitoring 在线质量监测
② fully oriented non-textured yarn 全取向非变形纱
③ FOY 全取向丝(纺丝一步法)
④ draw-wind, draw-twist or draw-warp 拉伸卷绕，拉伸加捻或拉伸整经
⑤ Co-We-Mat 青泽公司的激光监测自动落纱装置

ble to produce modified filament yarns on draw-twist and draw-wind machines. Differential shrinkage yarns are intermingled yarns of differing shrinkage. Differential-dye yarns comprise two or more yarns of differing dye affinity. Thin-thick yarns are a further possibility. An opportunity exists here for the textile processor to develop speciality yarns.

In warp yarns it is possible to use POY yarns and draw them by the warp-draw process. A few combined warp-draw-size machines[①] have been built but these have not met with wide success. Instead attempts are being made to bypass the sizing process for filament yarns by using a tangling process. It remains to be seen whether warp-draw will gain wide acceptance.

Twisting machinery

In twisting machines, in addition to effect twisters, two-for-one twisters and cabling machines enjoy continuing popularity. Cabling machines are used not for tyrecord and carpet yarns but also in lighter construction for sewing threads and embroidery yarns in viscose. Core twisting[②], yarn covering and air-jet covering are becoming more important for elastane yarns.

The Saurer "Tritec-Twister" is continuing to gain popularity. In twisting machines it is evident that speciality machines are superseding universal machines. Complete automation of twisting machines is not really economically viable. Meaningful and partial automation have been on offer in recent years. this is the route that has prospects of success.

---

① warp-draw-size machine 整经拉伸上浆联合机(WDS)
② core-twisting 包芯加捻

Vocabulary

discernible 可辨认的，看得清的
extent 范围
differential shrinkage 差异收缩量
comprise 包含，由……组成
instead（插入语）而，代之以
tangling 缠结
tyrecore 轮胎帘线
superseding 代替，取代，废除
partial 部分的，局部的

overall 全面的，普遍的
recommend 推荐，建议，介绍
modified filament 变性长丝，改性长丝
differential dye 差异染色
affinity 亲合性，亲合力
attempt 尝试，试验，试图
bypass 绕过，回避，越过
viscose 黏胶纤维
meaningful 有意义的

Think or answer these questions

1. What are the trends in texturing machinery?
2. How to produce fully oriented non-textured yarns?
3. How are differential shrinkage yarns composed?
4. How do differential-dye yarn be composed?
5. What yarns are cabling machines used for?

UNIT 5 MACHINERY FOR WOOL SPINNING 毛纺机械

LESSON 10
MACHINERY FOR WOOL SPINNING(1)

Nobody has expected technical or technological sensations in recent years as far as the worsted semi-worsted and woolen yarn industries were concerned. Rising costs plus international competition are forcing machine makers to consider their new developments from the aspects of economic viability and market acceptance within a reasonable time period. These factors are resulting in strategies in which tried-and-tested machine designs are being adaptively updated, justiable technological concessions are accepted for series production and also cost reductions are brought about by rationalization. Nevertheless several features of significance to the worsted semi-worsted and woolen spinning sectors are evident, namely

– The actuality and stability of tried-and tested technologies.

– Improvements of detail associated with machine cleanliness through improved air supply to machine elements.

– Improvements in drive and lubrication techniques.

– Optimization of dimensions of machine elements.

– Innovations in individual machine elements, though their practical significance remains to be proved by long-term trials.

– Improvement of data recording, analysis and storage.

– Improvement in machines for niche products.

– Improvements in handling and ergonomics[①].

Machines for worsted and semi-worsted spinning
– Carding

Effective machine width is restricted to 2500 mm. Productivity equals that of older worsted cards exceeding 3000 mm effective width due to the swift diameter being increased to 1500 mm plus higher peripheral speed. A new development is sliver delivery with double doffer and improved suction device. The new doffer comb provides up to 3600 strokes per minute. Remote adjustment of worker-to-swift clearance permits high precision setting, including following regrinding.

---

① improvement in handling and ergonomics 在操作及人类工程学方面的改进

– Combing machines

All combing machine makers have increased speed from 200 to 250~260 nips per minute. Further noteworthy innovations included:

The twin-head combing machine for first combing.

Web restriction and clean sliver formation thanks to adjustable blower action on leather apron and shovel plate.

An introduction was the "Rotat Feed Comb" rotating comber feed comb[①]. With this motion it is expected that cleanliness will be continuously maintained with less noise than the traditional feed motion.

– Drafters

There have been no significant improvements in worsted drafters. Leveler technology has progressed from the mechanical to the electronic. The associated benefits of easier programming, wider range of correction (±25%) and data recording are important for reducing lot-changing times, especially in the case of small batches. A very interesting development which reduces the number of mechanical drive elements is the intersecting with four frequency-controlled motors which incorporates mechanical sensing and electronic sliver mass control.

– High-draft rubber drawframes

The familiar high-draft rubber drawframes continue to be used with no significant changes.

– Worsted roving frames

The Bf rover with its very steeply angled drafting system has filled a technological gap[②]. The rovings delivered by the drafting units can be routed at the same angle and in the same length to the front and rear flyers. Roving tenions and twisting levels are constant even with delicate fibers.

– Worsted spinning frames

The "Plyfil 2000[③]" takes the form of a sophisticated spinning frame. Along with conventional refinements including automatic repair of thread breaks, automatic doffing

---

① "Rotat Feed Comb" rotating comber feed comb 旋转梳刷装置喂入梳栉
② ... filled a technological gap ……填补了一项技术空白
③ Plyfil 2000 绪森公司的 Plifil 2000 细纱机

with yarn length measurement, electronic thread clearers and linking process, the core-yarn attachment for Plifil is worthy of pamcular mention. Delivery rates as high as 280m/min can be attained. This technology is now fully under control.

Vocabulary

worsted 精纺的，精细的
woolen 粗纺的，粗梳的，羊毛的
strategy 对策，战略，策略
concession 让步，妥协
rationalization 合理化
actuality 现实，实际
cleanliness 清洁（度），净度
carding 梳毛
peripheral 外围的，圆周的
worker-to-swift clearance 梳毛辊（罗拉）相对于锡林的间隙
blower 吹风机
intersect 相交，和……交叉
steeply 陡峭地
roving 粗纱
thread clearer（电子）清纱器
semi-worsted 半精梳

reasonable 合理的，适当的
adaptively 适应地，适合地
justifiable 合理的，正当的
bring about 导致，造成
stability 稳定性），安定性
swift 大滚筒，大锡林
doffer comb 道夫斩刀
remote 远程，远距离
regrinding 碾磨
first combing 一级精梳毛
shovel plate（精梳机的）托持板
rover 三道粗纱机
gap 空白，间隙
delicate 脆弱的，易损的
core-yarn 包芯纱
as far as...be concerned 就……而论

Think or answer these questions

1. What are the significant features of the worsted. semi-worsted and woolen spinning sectors?
2. Does the effective machine width of the worsted cards be restricted?
3. What noteworthy innovations are the there in the combing machines?
4. Have there been significant improvements in worsted drafters?

5. Which ancillary devices do "Plifil 2000" include?

LESSON 11
MACHINERY FOR WOOL SPINNING(2)

– Ring spinning frames

The special feature of the "RKW" ring frame is the exceptional length of its top and bottom aprons. The bottom apron is driven by two bottom rollers with the top apron driven by frictional contact with two intermediate rollers. The name of this system is "Four-Roller-System with Toyota's① Original b CRADLE", Patent protection has been applied for.

Compact yarn spinning claims a large number of benefits② such as reduced hairiness, increased yarn strength, better fiber strength utilization, lower twist, higher spindle speeds, less fiber loss, higher productivity etc... problems in controlling the process need to be resolved by extended trials.

In the "Fiomax 2000"③ ring spinning frame. the use of HP-S 68 spindle bearings in association with a spindle whorl of only 17 mm diameter give it many advantage over other spindle drive systems. Energy savings as high as 20% can be achieved. Spindle speed variability is below 0.5%. The frequency-controlled drive④ is freely programmable and thus fully capable of controlled production of a very wide variety of qualities. The high-lights include doffing within a time of less than two minutes. One of the successful refinements with its simple design and mode of operation is the "Cut-Cat"⑤ whorl cleaning system. For stopping the spindle with the machine running the options are a stationary hand lever brake or a plug-in brake.

By using rotating ring⑥ in spinning yarns in the count range Nm 20 to Nm 60 it is claimed that production rates can be increased by up to 30%, end-breaks reduced,

---

① Toyota （日本）丰田公司
② Compact yarn spinning claims a lager number of benefits... 紧密纺纱工艺声称有许多益处……
③ Fiomsx 2000 绪森公司的 Fiomax 环锭纺纱机
④ frequency-controlled drive 频控传动
⑤ Cut-Cat Fiomax 2000 环锭细纱机上的自动锭盘清洁器
⑥ rotating ring 回转钢领

YARN AND FABRIC FORMING

现代纺织英语

traveler life extended to over 2000 hours and ring lubrication reduced.

– Ring frame drafting systems

One introduction was the PK 600 pneumatic weighting arm①.

Adjustment of the pneumatic weighting or the partial relief of pressure is performed centrally. The new UH 56 bottom apron cradle② makers it a simple matter to replace bottom aprons, thereby considerably reducing maintenance costs.

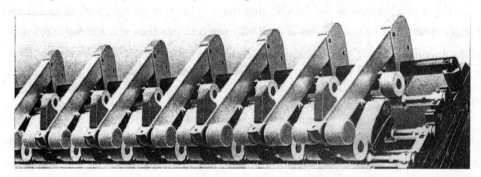

– Spindles

Although already well known, the benefits of the HP-S68 NSAS spindle③ were worth mentioning. Along with the virtually parallel course of bearing loading on the spindle collar with increasing spindle speeds, a development not yet full appreciated is the reduction of the noise level in all speed ranges by 5.5 ~ 6.0 dB (A) and the significantly lower vibration compared with spindles having rigid bearing.

– Ring/traveler systems

"Ceratwine"④ is the name given to a ring/traveler system with 100% ceramic ring and an appropriately

---

① PK 6000 pneumatic weighting arm　SKF 纺机配件公司开发的 PK 6000 气流加压摇臂
② UH 56 bottom apron cradle　UH 56 下皮圈托架
③ UP-S68 NASA spindle　绪森公司的 NASA(Noise Absorbing Spindle Assemble) HP-S68 型高速吸噪锭子
④ Ceratwine　（德国）Cerasive 公司的陶瓷钢领/钢丝圈系统的名称

designed ceramic-coated traveler. Its acceptance in worsted spinning needs to be supported by long-term trials plus analysis of its economic viability.

Semi-worsted spinning machinery

The OE 005 open-end rotor spinning machine① is designed for spinning counts in the range Nm 0.9 ~ 5.0. Its benefits are most appreciated when used in conjunction with a card sliver that has had one drafting process. With a rotor diameter of 150 mm rotor speeds up to 15000 rpm can realistically be attained.

Friction spinning with machines of the Dref 2 and Dref 3② types represents an alternative to semi-worsted and woolen ring spinning frames for counts up to Nm 10 for certain end-uses. The machines have been upgraded in certain details.

Semi-worsted ring spinning frames remain virtually unchanged. Sliver stop motions, semi-automatic or full automatic doffers plus data recording are now standard equipment. The high machine performance and relatively low demand mean that significant new models are unlikely to occur. Another factor is that in a major part of the market it has been superseded by the economically attractive wrap-spinning process.

The familiar "Robospin B6" mule spinner incorporating③ automatic doffer and condenser bobbin replacement is understandably used only for fine woolen yarns and high quality raw materials.

Improvements and updating have been in evidence over a wide spectrum in recent years, but frequently in detail only. These include:

– Improvements and modifications to drive systems.

– Improvements and simplification of lubrication systems.

– Improvements in drafter accessibility for maintenance and lot changing operations.

– Meaningful programmable systems for recording, storing and analysis with reproducible parameters for machine performance and settings, product quality and economic

---

① OE 005 open-end rotor spinning machine　OE 005 自由端气流纺纱机,系 NSC 公司(n. schlumberger & cie) 产品。国际上过去称为 open-end spinning(自由端纺纱),所以气流纱(国内俗称)叫 OE 纱。现在国际上规范称 rotor spinning(转杯纺纱)

② Dref 2 and Dref 3　Dref 2 和 Dref 3 型摩擦纺纱机,系奥地利 Fehrer 公司产品

③ Robospin" B6" mule spinning machine　Robospin B6 型走锭纺纱机,系 Bigali 公司产品

YARN AND FABRIC FORMING
现代纺织英语

viability.

— Improvement in pneumatic systems for machine cleaning.

— Abandonment of uneconomic refinements due to cost pressure①.

Some very valuable innovations were developed by suppliers to the machinery making industry, but their prospect of economic and practical success remain to be proved by extended industrial trials. These include:

— compact yarn spinning

— the pneumatic pendular arm

— the new UH 56 botom apron cradle

— the Rotat. Feed-Comb rotary feed comb for combing machines

— the "Ceratwine" ceramic ring/traveler system and

— the rotating ring

Contrary to what might be expected, it would be possible for the machinery makers to meet all needs of the yarn producer from the engineering aspect. But the fact is that with their speciality product ranges, what firms in this sector need is virtually universal machines or speciality machines and neither of these is economically viable. Shorter standing times and quicker lot changing have been achieved to a certain extent. But linking of processes continues to be impeded in terms of economic viability by technological and logistic constraints.

Vocabulary

exceptional 特殊的，异常的
extended 持续的，持久的
variability 变率，变化率
mode 模式，方式，形式
traveler 钢丝圈
parallel 平行
rigid 刚性的

open-end spinning 气流纺纱，自由端纺纱
card sliver 粗梳条子
remain 仍然，继续处于，保持
supersede 代替，取代，更换
condenser bobbin 搓条筒管
update 革新，不断改进
spectrum 领域，范围

---

① Abandonment of uneconomic refinements due to cost pressure 由于成本的压力，放弃不经济的改造

pressure 压力，艰辛
pendular 振动的，摆动的
intermediate 中间的
spindle whorl 锭盘
high-light 最精彩的地方，重点，显著部分
hand level 手柄杆，开关杆，调节杆
virtually 事实上
spindle collar 锭管
appropriately 适当地，恰当地
friction spinning 摩擦纺纱
realistically 实际的
occur 发生，出现，存在
mule spinner 走锭纺纱机
understandable 可以理解的

machine performance 机械状态
meaningful 有意义的
prospect 前景，预期，可能性
viable 能生存的，可行的，有生存能力的
standing 停止的，停着的，不在运动的
constraint 强迫，抑制，制约
be capable of 能……的，可以……的
be in evidence 明显，显而易见的
contrary to 和……相反
in association with... 与……结合
in conjunction with 与……协力，连同，连带着
be unlikely to 未必可能，不大可能

Think or answer these questions

1. Which benefits are there in compact yarn spinning?
2. Can the frequency-controlled drive be programmable?
3. What materials is "Ceratwine" ring/traveler system made from?
4. Which advantages do the HP-S68 NASA spindles have?
5. What interesting innovations were the developments by suppliers?

UNIT 6 AIR ENGINEERING 空气工程

LESSON 12
AIR ENGINEERING(1)

Air conditioning

Apart from the optimization of existing systems from the economic and performance aspects, there is a noticeable tendency for leading makers of air conditioning systems in the textile sector to move away from space air-conditioning① towards the dedicated conditioning of individual machines and process lines②. This emanates from changes in the air engineering③ requirements of modern high-performance machines. Due to heat source and airflows within the machine variability in temperature and humidity distribution can occur at sites critical to the process in question④ with consequent impairment of production and/or quality. Conventional air conditioning plant is no longer capable of eliminating this irregularity at reasonable cost.

Air filters

In the exhaust air filtration sector the drum-and-cartridge filter combining suction cleaning and waste disposal has become the accepted system. Most makers have made design modifications to enlarge the filtration area, reduce the overall size and thus significantly improve performance. Systems that may be mentioned here include Benninger⑤BENVAC, Luwa⑥Bahnson Multi Drum Vac, Wiesner, LTG⑦ Kompakt-Filter-System, Mazzini etc.

---

① space air conditioning　空间空气调节
② ...towards the dedicated conditioning of individual machines and process lines.　朝向专用的单独机器与生产线的调节
③ air engineering　空气工程
④ ...critical to the process in question...　对于相关工艺来说至关重要
⑤ Benninger　贝宁格公司
⑥ Luwa　鲁瓦公司
⑦ ...Benninger BENVAC, Luwa Bahnson Multi Drum Vas...　均为空气过滤系统名称, 如 BENVAC 为贝宁格直接整经机上的防尘系统

Air scrubbers[1]

Both Luwa and LTG[2] offered their new high-pressure scrubbers. The two systems incorporate nebulization of water at a pressure between 10 bar[3] and 100 bar. Droplet formation is substantially prevented and an aerosol is created, over 95% of which is re-sorbed by the airflow. This reduces the amount of water to be nebulized by over 99% compared with conventional scrubbers and the installed electrical power of the scrubber pump by about 80%. A further advantage of these scrubbers is that they do not operate with re-circulating water but exclusively with clean water.

This is a very welcome feature from the hygiene aspect. There is also no need to add biocides. But these scrubbers do demand higher water quality in terms of hardness and purity. To sum up it may be said that the higher cost of this new generation of scrubber is compensated by lower operating and maintenance cost.

Air control

In the air control sector there is a distinct trend towards dedicated and combination air conditioning systems as described earlier. Luwa (total air control[4]) and LTG (weave direct[5]) offered their systems fitted to weaving machines. In both systems the

---

① air scrubber 空气洗涤器
② LTG 总部设在德国 stuttgart 的 LTG 纺织空气工程公司
③ bar 气压单位(1bar = 0.9869 标准大气压)
④ total air control 全面空气控制
⑤ weave direct 织造定向的

air outlets are sited directly above the warp shed. With the warp yarn absorbing moisture very rapidly in the shedding zone, LTG experience is that warp end-breaks are reduced by 8%. The two systems differ in that with the Luwa system the conditioned air flows onto the warp shed (air-flow) whereas the LTG exerts a blowing action onto the shed with an air exchange rate of 60% to 80%. This also clears any lint from the warp. Both systems indirectly require air to be withdrawn though floor outlets underneath the warp. Adjustable blades in the LTG air outlet enable the airflow to be distributed in such a way that any uneven temperature distribution across the width of the warp can be compensated.

There is no corrosion of weaving machine or reed since the air outlets are automatically closed when the machine is stopped. Luwa also showed this system, on a ring spinning frame, with the air outlets being, sites above the creeled roving packages.

In the circular knitting sector Luwa developed an "air cap"[1] system. Special air outlets above the circular knitting machine cause an "air cap[2]" more than two metels in diameter to he formed around the machine. This prevents any external lint penetrating the cap. A homogeneous atmosphere is also created under the cap. Initial industrial trials suggest that the incidence of defects can be halved in circular knitting machines with protection of this type.

LTG adopts a different route with its "clean knit system". In this case the yarn feed packages are housed in an air-conditioned capsule[3]. The yarn packages are continuously kept free from lint by rotating air-jets[4]. The lint is suction extracted and returned with the exhaust air to the filter. In the knitting head zone the "clean knit system" uses a rotating air-jet which blows into an extraction duct any lint occurring as the needles are withdrawn. This lint is kept separate as it contaminated with oil residues.

---

[1] air cap 空气罩

[2] ... an "air cap" more than two meters in diameter to be formed around the machine. ……一个直径超过2米的"空气罩"在针织机周围形成

[3] ... the yarn feed packages are housed in an air-conditioned capsule. ……纱线喂入卷装被放置在一个空气调节的容器里

[4] rotating air-jet 旋转的喷气

Vocabulary

air conditioning 空气调节
move away 离去，退去
site 现场，工地，位置
impairment 损伤，减损，降低
irregularity 不规则，无规则
filter 过滤器
disposal 处理，控制，排出
dedicated 专用的
distribution 分布，分布状态，冲突
critical 苟刻，要求高的
consequent 由引…引起的，由…导致的
reasoning 合理的，适当的
exhaust 排气装置，排出的废气
air filtration 空气过滤
cartridge 滤筒，过滤器芯子
nebulization 喷雾（作用）
aerosol 气悬体，气溶胶
re-circulating 再循环，逆环流
outlet 出口，排出口
conditioned 调节的，调湿的
withdrawn 排出，去除
corrosion 腐蚀，侵蚀

external 外部的，外面的
homogeneous 均匀的
incidence 发生，影响，关联
halve 将……减半
extract 排出，抽出
needle 针，织针
overall 总的，全部的，综合的
droplet 微滴，小滴
resorb 再吸收，再重新吸收
sum up 总计，总结，概括
warp shed 梭口
exert 施加（力），运用，使受［产生］影响
underneath 在……下面，下面的
reed 筘，钢筘
penetrating 侵入，穿入
atmosphere 大气层
defect 疵点，织疵
house 容纳，放置，安放
duct 管，导管
contaminate 污染，弄脏

Think or answer these questions

1. What a noticeable tendency is there in air conditioning sector apart from the optimization of existing systems?
2. With what water do the new high-pressures scrubbers operate?
3. What a distinct trend is there in the air control sector?
4. Which advantages does the "air cap" system developed by Luwa have?

LESSON 13
AIR ENGINEERING(2)

Traveling cleaners

The cleaning sector is clearly dominated by conventional traveling cleaners such as those supplied by Luwa, LTG, Sohler, Neuenhauser①, Wiesner, Jacobi or Canalair. These systems have been updated, some fitted with new blower jets and equipped with additional electronic systems②. For instance the traveling cleaner hinges or rolls the suction or blower duct upward when it encounters an obstacle in its path③. This is an advantage in continuous cleaning, especially in cases where the track of the cleaner extends over whole lines of machines. Supplementary suction systems have also been incorporated, as with Sohler, for suction cleaning the drafting systems of the ring spinning frame.

In high-performance spinning and weaving machinery the trend with traveling cleaners is the purposeful suction extraction of lint directly where it occurs. Systems operating on the principle of forced removal by blower only are no longer capable of meeting requirements.

A totally new route in lint collection and elimination on ring spinning frames is being followed by LTG with its "clean spin system". This system has been developed specifically for modern high-performance ring spinning frames. It is aimed at collecting lint directly where it occurs. It utilizes the strong tangential airflow in the ring rail zone④.

---

① ...by Luwa, LTG, Sohler, Neuenhauser... by 后面为列出的公司名单, 如(德国)Sohler 公司, (德国) Neuenhauser 公司等, 它们提供适用于纺纱及织造加工部门的清洁装置

② ..., some fitted with new blower jets and equipped with electronic systems. ……, 一些配备新式吹风机喷嘴及装备附加的电子系统

③ the traveling cleaner hinges or rolls the suction or blower duct upward when it encounters an obstacle in its path. 当它遇到在行程中的一个障碍时, 巡回式清洁器用铰链转动或卷起吸风或低压空压机导管向上。

④ it utilities the strong tangential airflow in the ring rail zone. 该系统利用在环轨区内的强烈切向气流

By fitting a lint screen along the ring rail in spindle zone (attached to the doffer rail) the tangential airflow is directed through the screen. Any fibers carried in the air-stream are captured by the screen and are suction extracted by the traveling cleaner.

The fibers sucked from the lint screen are kept separate in the traveling cleaner and treated separately as this lint does not contain any extraneous fibers and be rerouted back into the production process[①]. LTG claims that in practical trials 22 kg lint per hour have been removed from 20000 spindles, whilst without screening it was only possible to remove 12 kg soiled lint from the floor. Since the screens can be hinged away, LTG reports little adverse effect on machine servicing.

To sum up it may be said that modern high-performance textile machines impose demands on air engineering and air conditioning technology[②] that conventional systems are no longer capable of fulfilling.

A strong innovation surge may be expected with the emphasis firstly on combination air-conditioning systems and secondly on cleaning, i.e. suction extraction of lint where it occurs.

Vocabulary

travelling cleaner 巡回清洁器
hinge 用铰链转动
dominate 支配，统治，占优势
encounter 遇到，碰到
obstacle 阻碍，障碍，干扰
supplementary 辅助的，附加的
follow 追求，追随，探索

adverse 不利的，有害的
fulfill 完成，满足，实现
track 轨道，轨迹
purposeful 有目的的
reroute 改线，绕行
impose 施加，利用，采用
surge 浪涌，急变，巨浪

---

① ... rerouted back into the production process ……改线返回到生产工序
② ... modern high-performance textile machines impose demands on air engineering and air conditioning technology... 现代高性能纺织机械对空气工程及空调技术提出要求……后面 that 引导的部分 demands 的定语

Think or answer these questions

1. In which aspects, do the traveling cleaners have heen updated?
2. For what machines is the "clean spin system" by LTG specifically?
3. What function does the lint screen have?
4. What drawback does the lint screen have?
5. What is the emphasis of innovation in air engineering?

UNIT 7 DATA PROCESSING 数据处理

LESSON 14
DATA PROCESSING(1)

Now it was evident that many machinery makers are once again placing the emphasis on engineering aspects. There were nevertheless several interesting advances towards greater adoption of networking within mills①. There were only tentative approaches to systems for combining technical data processing and administrative software.

The growing popularity of accreditation to quality standards of the ISO 9000 series②, including in the textile industry, has triggered off a trend in data processing to provide traceability in production within the various sectors. Both in process data collection and in production planning and control, valuable system proposals were offered by some companies.

In both these sectors of data processing there is a distinct trend to graphic interfaces and user-friendliness③.

---

① ... adoption of networking within mills.　……采纳工厂内部的联网
② ... accreditation to quality standards of the ISO 9000 series　……按照 ISO 9000 质量标准系列的鉴定, ISO (International Organization for Standardization 缩写)国际标准化组织
③ graphic interface and user-friendliness　图形界面与用户友好性

Production data collection in Yarn production

In the yarn production sector Rieter launched a data collection system to which its own machines can he linked in 1995. The system which is restricted to monitoring functions is attractive in its practicality and logic ally designed graphic interface.

Zellweger-Uster[1] too offered some products Along with a new combined capacitance/optical sensor there was a new software option which also enables nonperi odic thick spots to be recorded.

Another product was a ring spinning frame speed regulator controlled by the incidence of thread breakage, developed by LTV Denkendorf.

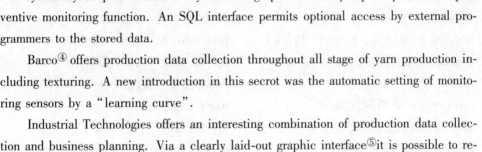

Outstanding features of the "Mill master[2]" system by Loepfe[3] include freely variable graphic summary reports plus a new preventive monitoring function. An SQL interface permits optional access by external programmers to the stored data.

Barco[4] offers production data collection throughout all stage of yarn production including texturing. A new introduction in this secrot was the automatic setting of monitoring sensors by a "learning curve".

Industrial Technologies offers an interesting combination of production data collection and business planning. Via a clearly laid-out graphic interface[5] it is possible to retrieve constantly updated production information. As distinct from many other systems,

---

[1] Zellweger-Uster （瑞士）蔡尔维格-乌斯特公司
[2] Mill-master 可与赐来福公司锭子识别结合的在线监控系统，由 Yarn-master 收集的质量数据，在 Yarn-master 中存储并整理
[3] Loepfe （瑞士）勒普菲公司
[4] Barco 在比利时 Kortrijk 的 Barco 公司
[5] SQL interface 结构化查询语言接口

it is also possible to retrieve comprehensive economic indices. A new feature in the system is a module for managing the raw materials inventory. By this means blends can be composed, future materials requirements can be managed, contracts drawn up and full financial analyses undertaken. The full system is also available for the Weaving sector.

Production data collection in fabric production

Banninger & hubscher①(a Zellweger associate company) developed several interesting technical innovations including a "pager" which like a personal message receiver is carried on the person and displays massagee. This function replaces the old central massage board. Likewise new is a warp production module② which besides displaying a planning board for warp preparation also permits speed and temperature gradients to be determined for the sizing machine.

Vocabulary

nevertheless 尽管如此，虽然……但是，然而
accreditation 鉴定
graphic 图形
capacitance 电容
tentative 试验性的，试用的，初步的
traceability 跟踪能力
restrict 限制
non-periodic 非周期性的
summary 简短的，概括地，累加的
learning curve 学习曲线
retrieve 收回，恢复，弥补
inventory 清单，报表，库存量

likewise 同样，也，又
gradient 增减率，变化率
contract draw up 拟定合同，起草合同
regulator 调节器，稳定器，控制器
freely 自由地，大量地
preventive 预防性的，防止的
lay-out 布局，展开，设计
index 指数，系数，指标
pager 记录器
besides 除……之外，除了
distinct from 和……不同
carry on 装在……上

---

① Banninge & Hubscher 瑞士一家 Zellweger 联合公司，产品为织造工厂专用软件
② warp production module 经纱生产模块

Tlhink or answer these questions

1. What is a distinct trend in data processing?
2. What are the outstanding features af the "Mill-master" system?
3. Can the "pager" be carried on the person?
4. Which stages of yarn production do the collection of production data include?
5. Can the warp production module be used for the speed and temperature to be determined on the sizing machine?

LESSON 15
DATA PROCESSING(2)

The new "Yarn Management①" module is noteworthy for having facilities for ongoing or "rolling" management of yarn stocks within a weaving plant. Using a twostage inventory code② it is also possible to record and reserve any yarn residues currently in production but not fully used up. The module provides a powerful means of tracing back any yarn used.

In addition to "Mill-master" system with its very clear practical graphic interface layout, Loepfe developed for the weaving sector a cloth inspection system of completely new design. The heart of the system is a touch-screen via which the person inspecting the cloth is able to undertake all the entries and retrievals needed for the task in hand. Neither keyboard nor mouse are needed③ to operate the system. All entries are designed in such a way that they can be made very easily on the touch-screen. This technique represents a pionegring step in the simplification of cloth inspection.

With its "Antara" system Incas④ offers very comprehensive process data collection software. New technical feature include an infrared remote controller for cloth inspection

---

① "Yarn Management" module "纱线管理"模块
② Two-stage inventory code 两级清单代码
③ Neither keyboard nor mouse are needed 既不需要键盘也不需要鼠标
④ Incas Incas 公司

and a NET-card which permits the rapid transfer of weave information. This high-speed transfer of information contributes additional benefits in quick-style-change.

A new feature offered by Barco for its PCMS system is a graphic interface. This system, suitable for weaving plants of small to medium size, is outstanding for its user-friendliness. For instance, list can be solted automatically column by column with clicking by the mouse①. The "Sycotex" system has now been equipped with several new elements and a graphic interface. Additional improvements are promised in the near future. Another new feature of the system is the facility for tracing right back to the yarn batch②. Complete terminal emulation of production data collection within the Picanol③Omni-Terminal has only been available for about a year. Hear if required it is possible to carry out a large number of retrieval operations on the machine terminal④which otherwise would only be possible on process data collection master -terminals⑤. It is thus possible to overlay a complete spares catalogue.

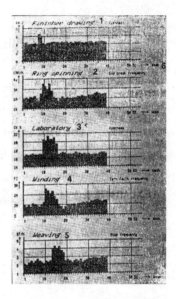

Production planning and control

The function spectrun of production planning and control software covers such diverse areas as order progressing, invoicing operations planning and materials management. In this field Datatex⑥ developed the "TIM" production planning and control programme that can be individually customized by table control. It has very comprehensive materials requirement and reservation functions plus built-in cost-accounting. An introduction was the programme package with graphic interface⑦. The already available pro-

---

① Lists can be sorted automatically column by column will clicking by the moues 随着鼠标的点击声，一揽表能够逐栏地被自动分类
② ...the facility for tracing right back to the yarn batch 容易正确地追溯纱线批料
③ Picanal （比利时）毕卡诺公司
④ carry out a large number of retrieval operations on the machine terminal 在机器终端执行大量检索操作
⑤ Ommi-terminal 全终端
⑥ Datatex Datatex 公司
⑦ ...the programme package with graphic interface 有图形界面的程序包

duction modules have been complemented with a new control station which enables project planning to be undertaken at individual machine level. The project plans can be run manually or using a variety of automatic planning techniques and revised at any time. The results are then transferred back into the "TIM" programme.

Practicability

Update developed the "Wespa" further upgraded production planning and control system with high practicability in the spinning, weaving and finishing sectors. The software package that can be run on AS/400 or UNIX① computers can be equipped with a variety of modules from marketing, purchasing, materials management and production sectors through to capacity and production planning. Another outstanding feature is the large number of interfaces with other task areas such as cost accounting, wage calculation, production data recording and many others. As far as the machine operator is concerned graphic interfaces are now state-of-the-art. The opportunities offered by this man/machine interface② for easily understood highly simple machine attendance whilst simultaneously minimizing the risk of error has recognized by machinery makers, and utilized although to differing degrees. As far as practicality is concerned there has been consolidation at the high level already attained.

Vocabulary

ongoing 正在进行着的，前进的，进行的   list 一览表，清单
code 代码   click 点击声
reserve 保存，储藏   promise 约定，答应，有……可能，预示
cloth 织物，布   emulation 模仿，仿真
retrieval 可取回，可收回，检索   overlay 重叠，叠加
pioneer 开辟，首创，走在前列   spare 备用的，备份的
cloth inspection 验布，织物检验   spectrum 领域，范围，系列
net-card 网络卡   invoice 开发票，开清单

---

① UNIX   UNIX 操作系统（交互式分时操作系统）
② man/machine interface   人/机界面

reservation 储备，备用
cost-accounting 成本会计
rolling 转动的，周而复始的，滚动的
record 记录，记载
trace back 回忆
touch-screen 触摸显示屏，触屏
mouse 多媒体计算机的鼠标
infrared 红外线
transfer 传递，传送，传输
column by column 逐栏地
sort 分类，区分
other wise 否则，在其它方面，此外
master 主要的，总的

built-in 内装的，嵌入的，固定的
project 规划，方案，项目
retrospect 回想，追溯的，追忆的
as far as... is concerned 就……而论
attendance 维护，看管，照料
recognize 认识到，承认，认可
complement 补充，互补，配套
revise 改变，修订，校正
capacity 生产能力
scrutiny 详尽研究，细看，仔细检查
state-of-the-art 现代化的，目前工艺水平的
simultaneously 同时地，同步地

Think or answer these questions

1. Which applications does the "Yarn Management" module have?
2. What is the heart of the cloth inspection system offered by Loepfe?
3. What is another new feature of the "Sycotex" system?
4. What is the function spectrum of production planning and control software?
5. With which modules can the "Wespa" software package be equipped?

UNIT 8 AUTOMATION OF HANDLING AND TRANSPORT 装卸与运输的自动化

LESSON 16
AUTOMATION OF HANDLING AND TRANSPORT(1)

Higher production rates

Production rates are still continuing to rise progressively in most production processes as a result of applied development work. Nevertheless, with certain exceptions, the rate of increase is declining because production speeds have already reached a very high level generally.

Under these circumstances, the time spent in materials handling, lot-changing and associated operations becomes increasingly important to the overall efficiency of production. Because of these interrelationships the need for inter-machine transport systems and automation of machine servicing[①]is more vital than ever.

Expensive automation

This is the reason way for many years now machine makers have been developing and suppling automated transport and handling systems. But it must not be over-looked that handling systems need to be compatible not only with the process in question but also with the space available and the production facilities of the particular site which demands additional project engineering and therefore generally involves relatively high cost.

Handling devices for machine feeding and product collection often need to perform complex movements and even undertake identification tasks otherwise performed by a machine operative. Efficient computing task can now be performed at very advantageous cost but the complex mechanisms of reliable handling devices are generally still very costly.

Cost depends on process

The outlay needed in any particular case depends very much on the process to be automated. In assessing whether or not automation is worthwhile there are other factors to be considered. however. These include the value of the materials processed, the wage structure of the plant, a very important factor, and aspects relating to quality assurance.

Because of these complex and constantly fluctuating inter-relationships it is very difficult for user, and the machine maker too, to predict whether a certain automation or

---

① inter-machine transport systems and automation of machine servicing　机器间的运输系统和机器维护的自动化

YARN AND FABRIC FORMING
现代纺织英语

handling device will or will not be acceptable. Evidence of this is provided by the fact that a number of automation devices, for example automatic handling systems for replenishing 2-for-1 twisters①, have disappeared since ITMA'91, whereas a quick-style-change system (QSC)② is now supplied by virtually every maker of weaving machines.

Can, lap and bobbin handling

A continuous can, lap and bobbin handling system for the spinning mill is also available from any major maker of spinning machinery. The devices often come from speciality makers of handling equipment such as Gualchierani, Neuenhauser Maschinenhau③ or U. T. L. T. A characteristic feature of these devices is that they are available in various levels of automation.

Roving bobbin handling

Some overhead rail conveyor systems④, for roving bobbins for example, are desiged in such a way that bobbin changing at the ring spinning frame can be performed manually with random creeling. On the other hand, the same maker supplies systems for block creeling, with the roving bobbins immediately being placed automatic aly in the feed position on the ring spinning frame⑤. All that is needed from the operative is to piece-up the feed sliver.

Can handling

In a number of spinning plants can handling from drawframes to speedframes already uses floorlevel track conveyors⑥. On the other hand driverless conveyor trucks for can handling were more strongly promoted(including for rectangular cans for rotor spinning⑦) by such firms as Rieter and Schlafhorst, and also for lap transport including systems by CSM and Vouk⑧.

---

① 2-For-1 twister　倍捻机

② quick-style-change system(QSC)　品种快速更换系统或快速换批系统

③ such as Gualchierani, Neuenhauser Maschinenhau... such as 后面为举例的公司, 如(意大利)U. T. I. T. 公司, 自1950年代起就从事工厂内部货物搬运设备的生产。

④ over head rail conveyor system　空中轨道运输系统

⑤ ... with the roving bobbins immediately being placed automatic aly in the feed position on the ring spinning frame ......使粗纱管立刻在环锭细纱机的喂入位置自动装入

⑥ floor level track conveyor　地面高度轨道运输线

⑦ rectangular cans for rotor spinning　转杯纺纱机用的矩形条筒

⑧ Vouk　(意大利)Vouk公司

Part 1 technical articles

课文

Vocabulary

rate 速度,速率
generally 广泛地,普遍地
circumstance 情况,情形,环境
vital 必需的,不可缺少,极其重要
compatible 相容的,兼容的,协调的
identification 识别,辨别,辨认
outlay 费用,经费,支出
worthwhile 值得做的,值得花时间的
predict 预测,预言
exception 例外,除外
decline 下降,减少
interrelationship 相互关系,互相影响,干扰
overlook 忽视,忽略,没注意到

advantageous 有利的,便利的
assessing 评价,评定
fluctuate 变动,变化,不定
replenish 补充,补给,再充满
can 条筒,圆筒
random 任意的,无规律的
piece-up 接头
floor-level 底板(层)
the need for 满足……的需要
characteristic 性能,特性
block creeling 分段换筒子,分段换条筒
conveyor 运输带,运输机,传送器
promote 提倡,促进,推销

LESSON 17
AUTOMATION OF HANDLING AND TRANSPORT(2)

General acceptance for Quick Style Change

In weaving at the present time the aspect of automation that is most in evidence is QSC. Almost every maker of weaving machines now supplies an appropriate system. In the demonstrations at ITMA'95 veritable "races" were even run[①] in which the time for a style change was displayed on a clock visible from far away. But direct comparison of style change times is difficult because conditions can vary greatly. It is interesting to note that the Genkinger[②] company supplies the basic truck used by five weaving machine makers for their QSC systems.

Progressive automation in weaving

In contrast to spinning which has numerous automated handling systems[③], automation of materials handling in weaving is still just beginning. At ITMA'91 Sulzer Ruti[④] and Innovatex[⑤] showed a weft package handling system with which weft packages were retrieved from store by an overhead conveyor and loaded into the creel of the weaving machine by a robot traveling on a craneway. The system was relinquished as it was not economically viable. At ITMA'95 a video film showed a handling system developed by Neuenhauser Maschinenbau in association with ITV Denkendorf. A traveling cleaner installation carries a grab system[⑥] which takes the weft packages from an overhead conveyor to specially designed creels with automatic loading on the weaving machines. This, along with numerous automation devices for the spinning plant, has facilities for progressive automation of handling, enabling it to be used rationally under a variety of production conditions.

Linked handling devices

The state-of-the-art is that it is possible to link handling systems to materials pro-

---

① veritable "races" were even run 甚至进行真正的"比赛"
② Genkinger （德国）Genkinger 公司
③ automated handling systems 自动装卸系统
④ Sulzer Ruti 苏尔寿·吕蒂公司
⑤ Innovatex 英诺维特克尔公司
⑥ A traveling cleaner installation carries a grab system... 一个巡回清洁器装置携带一个抓取系统

gressing and data processing systems①, merging them into a single unit and providing components of a computer integrated manufacturing system (CIM)②. But as CIM is outside the scope of this report it will not be discussed furthe③.

Economically viable solutions

Generally speaking numerous solutions now exist for performing complex transport and handling tasks. Whether these are economically viable depends on the particular case and needs to be very carefully analyzed.

On the other hand a solution offered by a machine maker can only be sustained if it can be sold sufficiently often with economic benefits. The market will decide which automation solutions will survive in the long term.

Vocabulary

appropriate 适当的，合适的，适应的
weft 纬，纬纱
relinquish 放弃，停止
rationally 理性上，合理地
video film 电视影片
merge...into... 把……合并成

veritable 真正的，名副其实的
craneway 吊车导轨
grab 抓取，夹子
sustain 承认，确认，准许
a variety of 种种，形形色色

Think or answer these questions

1. What is the aspect of automation that is most in evidence in weaving?
2. Which stage is the automation of materials handling in weaving?
3. Why was a weft package handling system developed by Sulzer Ruti and Innovatex relinquished?
4. What does the state-of-the art of handling systems mean?

---

① to link handling system into materials progressing and data processing system 把装卸系统连入原料进度和数据处理系统
② computer integrated manufacturing (CIM) 计算机集成制造
③ computer integrated manufacturing system (CIMS) 计算机集成制造系统

5. What will decide which automation solutions will survive?

UNIT 9 WEAVING PREPARATION 织造准备

LESSON 18
WEAVING PREPARATION

Thread tensioners/creels

In recent years several machinery makers have offered new developments in thread tensioners having low interference risk due to intermittent blower action or dust-tight encapsulation. New machine brakes were developed by Benninger and Karl Mayer[①].

Karl Mayer also offered an updated creel incorporating knotting and cutting. The Benninger stand included a very efficient suction extraction system and a crossed-end correction device.

Karl Mayer has developed a roll-off creel for split film. Two noteworthy innovations on this machine are a positively controlled pre-tension regulator and a presser roll which also acts as a tensioner roll.

Warping machines

Control of carriage advance has been perfected. Benninger has made a fully auto-

---

① Karl Mayer 卡尔·迈耶公司

matic warper and Karl Mayer a machine that has been substantially automated. Single-end sample warpers① have been newly developed or up graded. Creeling of multi-colored warps② has been fully automated by Sucker-Muller-Hacoba③ whilst Benninger offers an optically controlled creeling aid.

Beamers

There is nothing fundamentally new in beaming machines. There are now machines for 1400 mm flange diameter. Maximum beaming speed is 1400m/min. Benninger offers automatic creeling.

Sizing machines

Sizing technology remains unchanged. Nevertheless there were some noteworthy improvements. Individual drives with frequency-controlled three-phase motors, process data recording has been made simpler and more extensive, more companies than ever are offering liquor consumption measurement, and machine attendance has been made easier.

Machine control has been decentralized. Every module has its own stored programme control with bus-bar connection to the central computer④. Sucker-Muller-Hacoba continues to be the only supplier of a size application measurement and control system.

A noteworthy introduction was the first software system incorporated in a sizing machine, the "BEN-Size-Expert". It provides basic data specific to the warp being sizes for machine settings and recipe-related size application. Once the relevant data have been entered it also analyses and assesses the sizing and weaving performance.

Indigo dye-size machine⑤

The strong continuing growth in the denim market has encouraged Sucker-Muller-Hacoba and Benninger-Zell⑥ to develop new machines. There is an unmistakable trend

---

① Single-end sample warper 单纱试样整经机
② creeling of multi-colored warp 多色经纱的换筒
③ Sucker-Muller-Hacoba （德国）祖克-米勒-哈科巴公司
④ Every module has its own stored programme control with bus-bar connection to the central computer 每个模块具有自身带连到中央计算机上导线的储存程序控制器
⑤ indigo dye-size machine 靛蓝染色浆纱机
⑥ Benninger-Zell 贝宁格策尔公司

towards environmentally-friendly indigo dyeing①. The two companies offer different routes for minimizing the use of hydrosulphite.

Sucker-Muller-Hacoba uses a prevatted concentrate, avoids the risk of unfixed dye by reducing the liquor concentration to 2-3g/l and accelerates oxidation by "quick-oxidation". Benninger-Zell dyes and sets in a nitrogen atmosphere②, consequently a virtually oxygen-free dye liquor. With a new process control development for dyebaths, Benninger-Zell guarantees dyeing consistency and levelness along the full length of the batch and across the full width of the fabric. A first-time development was the "BEN-Link" system for batch changing without intermption of the dyeing process.

Vocabulary

interference 干涉，相互影响，干扰
dust-tight 防尘的
encapsulation 密封
crossed-end 绞经，经丝交叉，经丝交错
split film 裂膜
pre-tension 预张力，预加张力
single-end 单纱
optically 光学上的
beamer 轴经整经机
flange 边盘
three-phase 三相的
decentralize 分散
recipe-related 有关的制法，相关的处方
indigo 靛蓝，深紫蓝色
unmistakable 明显的，清楚明白的
intermittent 间歇的，中断的
knotting 打结，清除尾丝

cutting 切断
roll-off 滑离，辊轧
positively 确实地，必定
warping machine 整经机
multi-colored 多色的
aid 辅助设备，辅助手段
heaming machine 整经机，倒轴机
sizing machine 浆纱机
liquor 液，液体
bus-bar 汇流排，母线，工艺导线
enter 进入，引入
denim 牛仔布，粗斜纹布
hydrosulphite 亚硫酸氢盐
pre-vatted 预先放入大桶里
unfixed 不固色
nitrogen 氮
dyebath 染浴，染液

---

① environmentally-friendly indigo dyeing 环保的靛蓝染色
② ... dyes and sets in a nitrogen atmosphere ……在一种氮保护气氛中染色与固色

levelness 匀染性,水平度
concentrate 浓缩液,蒸浓液
quick-oxidation 快速氧化

consequently 从而,所以,必然
consistency 稠度,一致性
batch 一批,布卷

Think or answer these questions

1. Does the Karl Mayer stand include a suction extraction system?
2. Which one produced the full-automated warpers, Benninger or Karl Mayer?
3. Is there any new breakthrough in sizing technodogy sector?
4. For what use does the first software system incorporated in a sizing machine use?
5. Which route of the indigo dying dose Benninger-Zell adopt?

UNIT 10 SHED-FORMING MOTIONS 开口形成机构

LESSON 19
SHED-FORMING MOTIONS

The new development and updated systems in the shed-forming sector continue to follow the trend towards higher machine speeds, easier servicing and flexibility in fabric manufacture.

Weft insertion rates on weaving machines have been successfully further increased. The makers of shedding motions have responded with a new generation ot more stable machines of more compact design capable of coping with the higher speeds. The following details are worthy of particular note.

— The transmission elements between shed-forming motion and shaft now have centralized lubrication① and their linkages have been strengthened.

— Pulley mechanisms in jacquard motions are not only of larger and stronger design, they are also equipped with low-maintenance steel ball bearings②.

— The jacquard motion is detached from the body of the weaving machine and has independent synchronized drive③. This prevents any risk of inertia in long transmission mechanisms④.

— In high-speed electronically controlled jacquards there has been a change from upper-shed reading-in to lower-shed reading-in. By this means harness movement is smoother and the useful life of the motion can be considerably extended.

— In narrow-fabric weaving where machine speeds are generally particularly high, shaft return in negative motions can now be performed by pneumatic means⑤ and thus with greater elasticity.

— The diameter of harness return springs has been made narrower to reduce whiplash vibrations.

With regard to making fabric production more flexible some notable innovations can be reported. The maximum number of hooks that can be incorporated in a jacquard has been doubled or at least significantly increased. Some examples of monobloc systems are:

— 12288 hooks(Staubli⑥CX 1060)

— 10725 hooks (Grosse⑦ EJP-2)

— 13824 hooks comprising 24 jacquard modules each with 576 lifter elements (Tis⑧)

— 5120 three-position lifter elements for pile weaving(Staubli 1090)

---

① The transmission elements between shed-forming motion and shaft now have centralized lubrication... 开口形成机构与轴之间的传动部件现在已集中润滑……
② low-maintenance steel ball bearing 低维护钢球轴承
③ independent synchronized drive 单独同步传动
④ inertia in long transmission mechanisms 长传动中的惯性
⑤ ... shaft return in negative motion can now be pelformed by pneumatic means ……在消极装置中的回综现在能靠气流方式进行
⑥ Staubli （瑞士）斯陶布利公司，世界著名纺机公司之一
⑦ Grosse （德国）Grosse 公司
⑧ Tis （法国）Tis 公司，产品有电子提花机组件、打样织机等

This means that sets exceeding 60 threads/cm hecome possible in cloth widths around 180 cm if there is only one repeat across the full width. This permits any desired pattern to be woven without regard to junctions of repeats, including bordered and edged patterns, since every warp end can be individually controlled.

Of course energy consumption is greater with the larger number of magnets, but in some cases it has been considerably reduced.

The thread density in a jacquard motion can be modified by a double comber board or by a variable intermediate grid with the additional facility for shifting jacquard modules. The modular construction of electronic jacquard motions permits entire rows of modules to be added or removed in order to match the number of hooks to the particular repulrements.

Good progress has been made too in operational reliability and in the quality of fabric produced. This includes magnet monitoring in electronic jacquards①which can even take place with the machine running. Machine encapsulation prevents contamination by dust and permits forced ventilation②, shutters do not require any extra space above the jacquard.

The new generation of controllers leaves nothing to be desired. They are no longer regtricted with regard to repeat length. Funcrions have been extended and operation has become more user-friendly.

There is generally also compatibility with all current CAD③systems. Where required, corrections to pattern data can be undertaken quickly on-the-spot on the controller or also on the lap-top④. It is also now possible to produce simple jacquard drafts by hand-scanner and lap-top. Some weaving machines now incorporate mouse and IBM keyboard⑤.

There are innovations too in the weaving of heavy weight fabrics. A closed-shed dobby with electro-pneumatic control for a maximum of 24 shafts is available designed

---

① magnet monitoring in electronic jacquards   电子提花机中的电磁控制
② .. and permits forced ventilation   ……并允许受迫通风
③ CAD （computer aided design 的缩写）计算机辅助设计
④ corrections to pattern data can be undertaken quickly on-the-spot on the controller or alao on the lap-top   对花纹组织数据的修改，能当场在控制器上或也可在笔记本电脑上快速进行
⑤ IBM keyboard （International Business Machines Corporation）美国国际商用机器公司的健盘

for reed widths up to 3.5 meters.

For weavers not yet ready to replace their mechanically controlled jacquards, an electronic reading-in system is available for the Verdol needle motion.

The sensor needle is held or released by piezo-electronic means①. Retrofitting is possible on 1344-hook machines with connection to an electronic data processing system.

Vocabulary

shed-forming 开口形成
shedding motion 开口机构
shaft 轴
jacquard 提花机
upper-shed reading-in 梭口上层经纱读数
harness 综,提花机上的通丝
useful life 使用期限,有效期
weft insertion 引纬,导纬
transmission 传动,传递,传动装置
pulley 滑轮,皮带轮
detach 分解,分开,分离
inertia 惯性,惯量
lower-shed reading-in 梭口下层经纱读数
motion 运动机构,运动装置
narrow-fabric 带,狭幅织物
elasticity 弹性,灵活性
whiplash 抖动冲击
monobloc 单元的,单块的,整体的
three-position 双开口
set 织物经纬密度

full width 全幅
pattern 花纹组织,图案,花样
edge 给……加边
magnet 磁体,电磁体
grid 网格
ventilation 通风,换气
compatibility 兼容性,适用性
lap-top 笔记本电脑
heavyweight fabric 厚重织物
electro-pneumatic 电-气动的
shaft 综片
Verdol 韦多尔
retrofit 改型,式样翻新
return spring 回综弹簧
hook 提花机竖钩,多臂机拉钩
lifter 竖钩,拉钩,升降机构
pile weaving 天鹅绒织造
repeat 花纹循环,花纹完全组织
exceed 大于,超出,超过
bordered 加边

---

① The sensor needle is held or released by piezo-electronic means 借助压电方式传感器针被夹住或松开

junction 连接，接头，结
comber board 目板
encapsulation 密封，封闭
shutter 盖，挡风板
on-the-spot 现场的，当场的
hand-scanner 手持扫描仪

closed-shed 闭合开口
dobby 多臂机
not yet 尚未，还没有
piezo-electronic 压电的
respond with 报以，以……表示回答

Think or answer these questions

1. How have the makers of shedding motions responded to further increased weft insertion rates?
2. What advantage dose that the jacquard motion is detached from the body of the weaving machine has?
3. Why has diameter of harness spring been made narrower?
4. What drawback does the magnet monitoring in electronic jacquards has?
5. Can the correction to pattern data be undertaken on the lap-top?

UNIT 11 DEVELOPMENT TRENDS IN WEAVING
织造的发展趋势

LESSON 20
DEVELOPMENT TRENDS IN WEAVING(1)

After along gap there has once again been a genuine innovation in the weaving sector, the Sulzer Ruti "M8300" multi-phase weaving machine①. This machine, quite simply the sensation of ITMA'95, sagnaled a new era in weaving technology after thousands of years by the unchanged basic principle of single-phase weft insertion②.

---

① Sulzer Ruti "M8300" multi-phase weaving machine 苏尔寿·吕蒂 M8300 多梭口织机，该种织机依据的是多重线性开口原则，在织造滚筒控制下完成引纬，它包含了织造方法的革命，具备织造技术前沿性
② basic principle of single-phase weft insertion 单梭口引纬基本原理

YARN AND FABRIC FORMING

The question that naturally arises is whether this new technology can ever become successfully established because once before there was multiphase euphoria with ripple-shedding① which eventually faded into obscurity②. The prospects for the new system appear to be not too bad. Who in such a short time would have expected such a highly developed weaving system to be launched? However, we will first describe the most significant advances made in conventional weaving.

General developments in weaving machines
– Drive, control and computer technology

The real revolution in weaving machine technology over recent years has emanated from drive, control and computer technology. There has been a fundamental change in drive technology. Electronically controlled individual drives represent the basis for a computer-controlled weaving machine. Frequency-controlled motors or D. C. motors provide for speed changes of a major③ order, adapted to yarn and fabric produced. Drives with gear wheels are largely redundant④. Warp beam and cloth beam are electronically controlled independently.

Weft selection, weft accumulator, dobby and jacquard today have electronic con-

---

① multiphase euphoria with ripple-shedding 波形段口引起的多相兴奋
② faded into obscurity 消失在黑暗之中
③ Frequency-controlled motor or D. C. motors provide for speed changes of a major... 频控电机或直流电机提供了大数量级的速度变换……
④ Drives with gear wheels are largely redundant 齿轮传动基本上是多余的

trol. In this way it is possible by computer to change weft density and warp tension, weft system and weave, selectively including whilst the machine is running.

Sophisticated computer-controlled start-up procedures permit restarting without starting marks in virtually all woven fabrics[①]. A whole series of functions contribute towards this: shaft-leveling, removal of warp tension at stop, start-up correction (tension), overspeed in accelerating, automatic miss-pick motion, automatic repair of weft breaks, switching to another weft accumulator, beat-up control and double-ended slay drive.

All these electronically controlled weaving machine motions developed over recent years today from the basis for the fully electronic computer-controlled weaving machine with substantially simplified operation.

– Technological improvements

The new drive and control techniques have also made technological improvements possible. For instance controlled weft tesioners enable weft tension to be made compatible with the yarn concerned and varied over the course of time, during the weft insertion cycle. This is advantageous in all weaving systems such as projectile, rapier and air-jet machines. In the case of air-jet weaving machines for example, the high peak tension occurring when the welf thread stops abruptly and which restricts the rate of acceleration can be reduced to half. This reduces stress on the welf yarn and offers the opportunity to increase further the speed of the air-jet weaving machine.

Weft presenter systems with electronically controlled individual drive[②] permit positive positioning of the feeder, which immediately the weft thread has been prevented is returned to a convenient operating position to prevent any unnecessary deflection points in the yarn path.

Weft accumulators have likewise been further improved. Controlled let-off tensioners with a compensating action[③] can be momentarily opened by electronic means. One weft accumulator was fitted with a device for humidifying or applying additive to the weft

---

① ... permit restarting without stating marks in virtually all woven fabrics ……程序允许重新启动，而不需要在所有的机织物上标明标记
② Weft presenter systems with electronically controlled individual drive 电子控制单独传动的引纬器系统
③ controlled let-off tensioners with a compensating action... 有补偿作用的受控送经张力器……

yarn. Positively controlled back-rests, some with considerably reduced dimensions, are capable of positively varyiag the dynamic course of warp end tension so that, for example, a cleaner quicker shed opening is obtained. This means less stress on the warp ends, lower mean warp thread tension and the possibility of higher speed. One area where electronics have not yet penetrated is in the adjustment of the dimensions of the shed①. Correct shed configuration is essefltial to smooth weaving machine performance. Unfortunately there are just a few weaving machines, for example the rapier machines "G6200" by Sulzer Ruti or the "FAST" by Nuovo Pignone②, which provide the weaver with facilities for precisely reproducible transferable setting of shed dimensions.

Vocabulary

gap 中断，间断
multi-phase weaving machine 多相织机，多梭口织机
euphoria 幸福，欣快感
eventually 终于，最后
basis 基础
redundant 多余的，重复的
weave 组织，织纹
start-up 起动，开动
starting marks 开车痕
weft break 断纬
slay 筘座
presenter 储纬器，纱线引出器
abruptly 突然地
positioning 定位，配置，位置控制
deflection 偏离，偏转
momentarily 一瞬间

shed opening 开口高度
contribute towards 有助于，为……出力
in the case of 就……来说，至于
essential to 对……是必不可少的
genuine 真正的，纯粹的
signal 成为预兆
arise 产生，出现
ripple-shedding 波形梭口
obscurity 黑暗，黑暗处
gear wheel 齿轮
cloth beam 卷布辊
sophisticated 复杂的，成熟的，完善
procedure 程序，方法
woven fabric 机织织物
beat up 打纬
weft accumulator 无梭织机的储纬器
peak 波峰，顶点，引入脉冲点

① One area where electronics have not yet penerated is... 电子仪器还没有渗入的一个区域是……
② Nuovo Pignone 新比隆公司

weft thread 纬纱线
convenient 合适的
let-off tensioner 送经张力器

back res(织机)的后梁
penetrate 进入，渗入

Think or answer these questions

1. Why is the M8300 multi-phase weaving machine said to be a genuine innovation in the weaving sector?
2. Which aspects are the general developments in weaving machine embodied in?
3. What advantages has the controlled weft tensionerd?
4. What advantages has the controlled back rests?
5. What is the one area where electronics have not yet penetrated?

LESSON 21
DEVELOPMENT TRENDS IN WEAVING (2)

Developments in conventional weaving machines
– Projectile weaving machine

Running counter to the trend towards weaving machines with full electronic control and equipped with all manner of refinements, which naturally is far from cheap, there is increasing development of simple equipped standard weaving machines which are able to weave at lower cost. One example is the new "P lean" projectile weaving machine by Sulzer Ruti. In projectile weaving machines weft insertion rates of around 1400m/min have been attained with appropriate reed width (3.92 m, 400 rpm) and even as high as 1500m/min. projectile weaving machines are available in special width up to 8.5 m.

– Rapier weaving machines

There is still further development potential in rapier weaving machines. For instance the design of the rapier head is progressively being improved. Smaller lighter rapier heads permit higher speeds and lower shed heights. This means lower stress on the warp ends. Weft insertion rates on rapier weaving machines has conclusively gone beyond 1000m/min. the "floating" rapier①, which does not require any guide elements projecting into the shed, brings advantages in the splitting of the shed plus less damage to the warp ends.

Picanal has now applied the quick-style-change concept of the air-jet weaving machine to its "GTX" rapier weaving machine. As in the "Omni" and "Delta" types, the entire warp module can now be changed on the "GTX" also.

ICBT②, the well-known maker of texturing machines, has taken over the original Saurer-Diedrichs machine from Vamatex③ and affers two machines with controlled rapiers. As with Picanal, the machines were equipped with a warp module which is also self-mobile, in other words it does not need any other means of transport. This should be attractive from the aspect of cost. ICBT has introduced a new jacquard system in which the jacquard frame forms an integral part of the weaving machine frame. The result is an astonishingly low structure with advantages in respect of vibration.

Nuovo Pignone with its "NEXT" weaving machine presented a prototype based on the "FAST" series, with futuristically styled cladding intended to make a significant contribution to noise reduction. Somet④ also offered machine cladding incorporating noise reduction which is claimed to reduce noise by 2 dB (A). Somet also had a new

---

① "floating" rapier "漂移"的剑杆
② ICBT （法国）ICBT集团公司，位于法国 Rhone Alpes 地区
③ Vamatex （意大利）范美德公司，剑杆织机制造商
④ Somet （意大利）舒美特公司，该公司在剑杆织机市场上居先导地位

combined weft clamp and cutter which reduces weft waste at the weft entry side without the aid of a tuck selvedge. Another maker showing a special weft clamp for reducing weft waste was Sulzer Ruti.

On the "G6200" it reduces weft waste at the left side by 4 cm. Anyone using expensive weft yarns can immediately appreciate the potential payback. Sulzer Ruti developed a dynamic pile height control motion on a "G6200" terry machine which opens up novel design opportunities.

The automation of weft break repair on rapier weaving machines, which is relatively costly to undertake, was offered only by Somet and Vamatex. Somet employs a mechanical principle whilst Vamatex uses a pneumatic system.

The makers of rapier weaving machines demonstrated the high versatility of their machines with a multitude of speciality fabrics using the finest and coarsest yarns. For example, Dornier① offered "on-the-move" design changing② which performed a style change without stop on a rapier weaving machine③. The same warp was used of course but four new weft yarns were inserted and weft density, speed, warp tension and weave were changed.

Carpet weaving machines operating on the rapier insertion principle must also be mentioned. Here Van de Wiele④ developed the first rigid-rapier weaving machine to have three rapiers. By simultaneously inserting three picks, fifty percent higher productivity is achieved at the same speed. It is also possible to increase the speed.

– Air-jet weaving machines

The makers of air-jet weaving machines produced proof that they too can now offer not only productivity but also the ability to weave difficult yarns and produce thick heavyweight fabric. High-twist crepe yarns, chenille and delicate wool and filament weft yarns were being woven at high speed, also denim at high productivity and jacquard designs at over 1000 rpm.

Dornier demonstrated the manufacture of terry with chenille borders, though in this

---

① Dornier （德国）道尼尔公司, 剑杆织机制造离
② "on-the-move" design changing "在移动中"的设计变化
③ a style change without stop on a rapier weaving machine 在剑杆织机上不用停机而改变风格
④ Van de Wiele （比利时）Van de Wiele 公司, 其产品适于织造地毯和天鹅绒

case the speed had to be reduced momentarily for inserting the chenille wefe.

Gunne①, the fist maker to offer a teny air-jet weaving machine several years ago, provided a new frequency-controlled machine with low-vibration machine frame.②

Picanal demonstrated high flexibility at high speed with its air-jet weaving machines. For instance on an "Omni" a high-twist crepe yarn was being inserted with the aid of a new "extender" jet. A furnishing fabric with chenille weft was being woven at 750 rpm. Whereas the "Omni" is desiged for flexibility, the low-cost "Delta" machine is designed specifically for the production of standard fabric at more reasonable cost. Picanol also has developed a weft over-run sensor.

Vocabulary

refinement 经过改进的装置
rapier head 剑杆头
self-mobile 自动的
futuristically 未来主义地
weft clamp and cutter 纬纱夹与剪纬刀
versatility 多方面性,多方面适应性
coarsest 最粗的
weft density 纬密
rigid-rapier weaving machine 刚性剑杆织机
crepe yarn 绉纱线
whereas 而,却,其实
run counter to 和……背道而驰,违反……
weft insertion rate 引纬率

conclusively 确实,断然
in other word 换句话说,换言之
prototype 典型,范例
cladding 涂层,(金属)包层
terry 毛圈织物,起毛毛圈
a multitude of 许多的
on-the-move 在进展中,在移动中
carpet weaving machine 地毯织机
pick 纬纱,投梭
chenille 绳绒线,仿绳绒线
over-run 超过正常范围,超程
far from 远非,远离,并不是
all manner of 各种各样的
as with 正如……情况一样

---

① Gunne （德国）居内公司,专用织机(如天鹅绒剑杆织机)制造商
② a new frequency-controlled machine with low-vibration machine frame 一种有低振动机架的新型频控机器

Think or answer these questions

1. What is the purpose of the development of simply equipped standard weaving machines?
2. Why is the rapier head being improved to smaller lighter ones?
3. What does the cladding is intended for?
5. Which principle does Somet employ to implement the automation of weft break repair?
6. Is it possible to perform a style change without stop on a rapier weaving machine?

LESSON 22
DEVELOPMENT TRENDS IN WEAVING(3)

A noteworthy prototype is the douhle-sided "MACH3" air-jet weaving machine by Somet. This a hybrid machine with one drive. With the vertical arrangement of the warp path-warp beam below and cloth beam above①- the double machine occupies jus the same amount of space as a single machine. With two cloth widths each of 3.22 m and operating at 600 rpm, a weft insertion rate of 3864m/min is achieved, the highest rate of weft insertion of any air-jet weaving machine at ITMA'95.

Tsudakoma② developed two types of machines. The "ZAX" is intended for the production of standard fabrics at maximum speeds. A 1.7 meter reed-width machine of this type weaving cotton goods was running at 1700 rpm and the weft insertion rate of 2890m/min was the highest on a single-width machine. But the strengths of the "ZA 209i" lie in its flexibility. ③ For terry fabrics there is the "ZA 207i". The "AFR" universal fully automatic weft break repair system④ is also able to repair weft breaks before the main jet.

---

① warp beam blow and cloth beam above 经轴在下与卷布辊在上
② Tsudakoma （日本）津田驹公司，高速织机制造商
③ ZA 209i、ZA 207i 津田驹 ZA 209i、ZA 207i 型喷气织机，其中"i"代表智能化型，它们采用了智能键盘 i-Board
④ "AFR" universal fully automatic weft break repair sysrem AFR 通用全自动断纬修复系统

– Water-jet weaving machines

Water-jet weaving machines are little used in Europe. In endeavors to weave filament warps without sizing, serious thought should again be given to employing this weaving machine. With the absence of relay jets or weft control dogs these machines are ideal for weaving unsized air textured filament yarns.

– Dobbies and jacquards

The flexibility strived for in air-jet weaving machine places severe demands on the makers of dobbies and jacquards to make available correspondingly efficient units capable of high-speed operation. Electronic dobbies reach around 1000 rpm if only a few shafts are used. With sixteen shafts the attainable speed is only about 750 rpm. Cam-shedding machine get up to 1100 rpm. Rotary dobbies run at 500 to 600 rpm or somewhat less when using twenty or more shafts.

An interesting though not yet full commercial innovation was developed by Tsudakoma, namely individual drive to every shaft[1]. By this means the movement cycle and timing of shed closure can be controlled independently on every shaft. The new "Optifil" heald by Grob[2] is worth a mention, it promises a smoother passage for the yarn.

The situation in jacquards is especially difficult where harness involves spring-return elements. In recent years, high-speed jacquards running at speeds in excess of

---

[1] individual drive to every shaft 单独驱动每片综
[2] Grob （瑞士）格罗公司

1000 rpm have developed by Bonas① and Staubli. In practice it might be thought that such high speeds and the smooth running required would be difficult to achieve at present, since it was just a few years ago when the electronic iacquard was introduced that a speed as low as 400 rpm was considered to be a great achievement. Today the jacquard plus harness is expected to make the jump to 1000 rpm in the shortest of times. Development on the still slow-running monobloc jacquard with single-end control-Grosse and Staubli offered jacquards with more than 10000 hooks-is still aggravated by problems of harness wear because the loading on the hook must be applied by a single harness cord and its return-spring via increased tension②.

Experiments are already under way with independent drive for the jacquard.

Schleicher③ reads-in the hooks with a bi-metal strip, in other words without magnets. This system has low energy consumption and consequently generates less heat. Up to the present time this system is only available as a replacement part for the needle grid.

A flexible jacquard of modular design operating on the single-lift principle and without pulley lifting system was developed by Tis. Each module controls a maximum of 576 ends and is slide-mounted on a basic frame. A special comber hoard enables the warp density to be varied④. Alignment of the individual modules enables the harness motion to be substantially vertical.

– Weaving machine environment

In weaving, environmental conditions, especially relative humidity, have a major influence on the success of the weaving process. There has been recent success in improving dust removal from the weaving machine with individual air conditioning. With spun yarn warps it is advantageous to have a system with a forced downward airflow in the rear shed area of the machine with the downward flow of conditioned air assisting the deposition of lint on the floor. Lint is thus immediately carried downward away from the

---

① Bonas　在英国 Gateshhead 的博纳斯机器公司
② the loading on the hook must he applied by a single harness cord and its return-spring via increased tension　竖钩上的载荷必须由单个通丝和其回综弹簧通过增加的张力来施加
③ Schleicher　（德国）Schleicher 公司
④ A special comber board enables the warp density to be varied　一种特殊的目板能使经纱密度变化

work area. The extraction point for dust-laden air is best sited underneath the rear shed because this is where most fiber abrasion loss occurs. Systems of this kind, as supplied by LTG and Luwa, improve conditions in the work area and result in better performance. Direct machine air-conditioning is also the most economical system①.

Vocabulary

hybrid 混合的，杂化的
single width 单幅
strength 力量，力气；实力
endeavors to 力图，争取，努力
dogs 挡块，掣子
severe 急剧的，猛烈的；严重的
aggravate 加重，加剧，使恶化
bi-mental strip 双金属条
slide-mounted 滑动安装的
reed width 筘幅，在筘经幅
cotton goods 棉织品，棉制品
water-jet weaving machine 喷水织机
relay jets 接力喷嘴

unsized 无浆
commercial 大批生产的，商品化的
return spring 回综弹簧
up to the present 到目前为止，至今
comber board 提花目板
relative humidity 相对湿度
dust-laden air 带尘空气
strive for 争取，为……而努力
be expected to 可料到，可预期，应该
spun yarn 短纤纱
abrasion 擦伤，磨损，磨耗
get up to 到达，赶上

Think or answer these questions

1. Which uses are the two types of air-jet machines developed by Tsudakoma intended separately for?
2. Are the water-jet weaving machines largely used in Europe?
3. How fast are the speeds of high-speed jacquards developed by Bonas and Staubli?
4. What principle does the jacquard developed by Tis operate on?
5. What place is the extraction point for dust-laden air sited on?

---

① Direct machine air-conditioning is also the most economical system 直接的机器空调也是最经济的系统

LESSON 23
DEVELOPMENT TRENDS IN WEAVING(4)

- **Style-change and warp-change**

In recent years automatic cloth beam changing system have been developed by allmost all makers of weaving machines. Automation at this level is in fact state-of-the art but it must always be weighed against the argument whether a batched roll with its long piece length and opportunity for fabric inspection is not more cost-effective.

Warp-change and style-change are particularly interesting from the economic aspect. Today quick-style-change is offered by virtually all weaving machine makers, most employing a Genkinger quick-style-change trolley①. A trolley of this kind is now capable of aligning itself paralled to the weaving machine, making quick loading into the machine possible. There is a new quick-style-change trolley with facilities for loading two complete warp beams simultaneously. Handling has been further improved with the development of quick-fix heald fasteners②.

The Picanol quick-style-change system differs from those of other makers in that it has a module that can be removed from the weaving machine. This enables wraps preparation to be substantially completed in the drawing-in department. Picanol also developed a quick-warp-change system based on the style-change system, with warp tying being performed in the drawing-in department.

The multi-phase weaving system

In the Sulzer Ruti multi-phase shedding system several parallel sheds are formed which move in the warps direction. The "M8300" employs a "weaving rotor③ which has a number of shed forming systems. The warp partially around the rotor. It is thus possible to form four sheds in series④. Because of the small diameter of the rotor the height of the shed is very low and frictional contact areas short. Shedding is accomplished by guide bars which by means of short lateral movements determine whether the warp is

---

① Genkinger quality-style-change trolley　Genkinger 公司的快速换批小车
② ...with the development of quick-fix heald fasteners　……随着快速安装结合件的开发
③ "M8300" employs a "weaving rotor"　"M8300"采用了"织造的旋转件"
④ to form four sheds in series　连续形成四个开口

placed on or adjacent to the shed-forming element. The shed-forming elements incorporate an air channel by which means the weft thread is inserted. Weft is inserted into the four sheds in timed succession by means of air-jets with very low air pressure. The "M8300" is equipped with a number of microprocessors and has full electronic control①. It has automatic weft selection and automatic repair of weft breaks. The machine operates from warp beams of up to 1.6 meter. It is already provided with a quick-style-change system also capable of changing the overhead mounted cloth beam②.

The "M8300" is already at an advanced state of development. As distinct from the 1970s machines, this machine could in fact represent the start of a new era in weaving technology. First of all it already has extremely high performance, namely a weft insertion rate exceeding 5000 meters per minute plus apparently considerable development potential, and secondly this new development is backed by a powerful weaving machine maker③, again in contrast to the 1970s dvelopments. This is of the utmost importance in such a complex new development in which virtually nothing could be imported from conventional weaving technology, but all essential functional elements had to be designed right from basics④. The prospects for this new weaving system can therefore be regarded as good.

Weft insertion rates

The diagram (fig. 1) illustrates the various maximum weft insertion rate demonstrated at ITMA. Rapier weaving machine have without exception passed the 1000 m/min mark. Projectile machines are now reaching 1500 m/min. A single-width air-jet machine was weaving at a weft insertion rate of 2890 m/min and the Somet doublewidth machine even at 3864 m/min. if we look back just three decades⑤ it is clear meteoric progress has been achieved in conventional singphase weaving.

---

① full electronic control　全电子控制机构
② also capale of changing the overhead mounted cloth beam　也能更换安装在机顶上的卷布辊
③ ...is backed by a powerful weaving machine maker　……由强有力的织机制造商做后盾
④ …but all essential functional elements had to be designed right from basics　……但所有重要的功能元件都必需从基础而设计出发
⑤ if we look back just three decades...　如果我们追溯到30年以前……

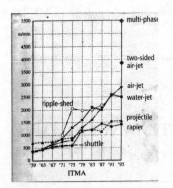

The pinnacle, in the form of the performance of the multi-phase weaving machine with its weft insertion rate of 5500m/min, marks the dawn of a new era① in weaving technology.

Vocabulary

weigh against 考虑，权衡，对比
piece length 匹长
lateral 侧向的，横向的
illustrate 图解，说明
look back 回顾，追溯，回头看
pinnacle 尖峰，顶峰，极点
adjacent to 靠近，接近，与……邻接
batched roll 布卷

guide bar 导纱梳栉
determine 测定，判断
diagram 曲线图，图表，图
without exception 无例外地，毫无例外
meteoric 迅速的，流星的
parallel to 与……平行
distinct from 和……不同

Think or answer these questions

1. In what place does the Picanol QSC system differ from those of other makers?
2. With which systems does the "M8300" be equipped?
3. Which weaving machine has the highest weft insertion rate?

---

① dawn of a new era 新世纪的开端

4. Are there any thing imported from conventional weaving technology in M8300 multi-phase weaving machine?

5. How much is the weft insertion rate of air-jet weaving machines now reaching?

UNIT 12 MACHINES FOR CARPET MANUFACTURE
制造地毯的机器

LESSON 24
MACHINES FOR CARPET MANUFACTURE

Networking of carpet machines by computer technology has been further perfected. Benefits include still higher carpet quality and productivity, shorter style-change times and more.

Tufting machines

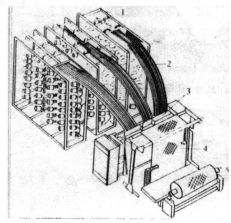

The "TX 300" tufting machine was offered by Tuflex①. Microprocessors directly control the machine functions using servomotors②. This keeps moving masses low and high production speeds are possible. This tufting machine incorporates two sliding needle bars for staggered pile, each being controlled by a servomotor. The designs to be produced are freely programmable by computer-controlled servomotors③. A major feature of the machine is the precision yarn feed for linear and staggered pile, with the yarrn being fed in an extended state.

The scissor-like cut involving gripper and knife has been superseded by a guillo-

---

① Tuflex （英国）Tuflex 公司

② Microprocessors directly control the machine functions using servomotor 微处理器直接控制应用伺服电机的机器功能

③ The design to be produced are freely programmable by computer-controlled servomotors 被生产的花纹靠由计算机控制的伺服电机可任意编程

tine-type blade movement. This features are aimed at eliminating shearing for cut-pile carpets, thereby economising on pile yearns.

Carpet weaving machines

The modern weaving plant is likely to feature face-to-face rapier weaving machines with electronic pattern control[1] which provides rapid pattern changes within a color palette determined by the available pile yarns. Pattern changing can be associated with problem-free pile weave alternation between single-cycle, two-cycle or three-cycle operation. When changing the weave of the chain warp and stuffer warp ends it is possible to change the shaft cam in iust a few hours. With computer aid weaving batches can be assembled by reed width and length to minimize cutting waste[2]. This is standard equipment on machines by Van de Wiele and Schonherr Chemnitzer Webmaschinenbau GmbH[3].

Whilst Schonherr developed the two-cycle two-shot weave free from indistinct definition in pattern and color, an innovation offered by Van de Wiele was traditional three-shot two-cycle weaving on a three-rapier machine. With this form of weaving which has been widely used since 1932 in three-cycle form. Carpets woven through to the back are produced with clear pattern definition. The dead pile is woven-in uniformly distributed between top and bottom carpets.

A new design of electronically controlled double-lift open-shed jacquard was developed by Staubli, the "CX 1090B". Eight pile colors were available for patterning. By a novel variation on a two-cycle two-shot weave woven through to the back, thirty-six different colorways including mixed colors were produced in the carpet[4], utilizing to the full the impressive versatility of the electronic patterning system.

Hemaks[5] offered a face-to-face carpet weaving machine with single flexible rapier

---

① The modern weaving plant is likely to feature face-to-face rapier weaving machines with electronic pattern control.... 现代织造设备似乎要以电子花型控制的双层绒头织物剑杆织机为特色……

② With computer aid weaving batches can be assembled by reed width and length to minimize cutting waste 由于计算机的帮助，织造的批量能按筘幅与长度配合以减少切断的浪费

③ Schonhen Cheranitzer Webmaschinenbau GmbH （德国）Schonherr 公司

④ thirty-six different colorways including mixed colors were produced in the carpet 36 种不同的色位包括混合色在地毯上产生

⑤ Hemaks （土尔其）Hemaks 公司

YARN AND FABRIC FORMING
现代纺织英语

of very simple design for the productioa of carpets with scraped back① and machine with two flexible rapiers for traditional three-cycle three-shot weave carpets. Vedol jacquards along with harness and back-scraper motion are built by the company. This machine is used predominantly by small-scale producer in Turkey and Middle East②.

To reduce machine cleaning time, makers of carpet weaving machines should think about the possibility of supplying auxiliary dust extraction systems for use in the weft insertion and/or shaft aress③. These would be advantageous when weaving jute and/or wool yarns.

Carpet jacquards

In jacquard engineering, a significant innovation by some suppliers is to replace the magnets used for electronic pattern control with piezoelectric elements in special ceramics④. These act as flexural resonators. The piezo elements possess the following advantages: small dimensions of the dust-tight encapsulated cassettes or modules, low power consumption and absence of any need for cooling. Two half-hook are replaced by one full hook. Schlecher and Takemura⑤ offer retrofit sets incorporating piezo elements for all mechanically controlled jacquards for carpet weaving and flat weaving.

They replace Verdol and French fine-pitch jacquard cardsa⑥. Piezo elements enable energy consumption to be reduced by more than 95% compared with electro-magnet control. Takemura transmits pattern information by light guide. One piezo module for 28-needle Verdol is just 5 mm thick and only fifty Watts are needed to drive 1344 hooks.

Karl Mayer is the fist company to offer a piezo-element based zero-harness jacquard⑦ for its curtaining machine. Various suppliers have upgraded their magnet-control

---

① for the production of carpets with scraped back   适合有摩擦背面地毯的生产
② in the Turkey and the Middle East   在土尔其及中东
③ ... supplyink auxiliary dust extraction systems for use in the weft insertion and/ or shaft areaa   ……提供适于用在引纬和/或综片区域的附加除尘系统
④ ... to replace the magnets used for electronic pattern control with piezoelectric elements in special ceramics   ……用特殊淘瓷压电元件取代用于电子花纹控制的磁件
⑤ Takemura （日本）Takemura 公司
⑥ Verdol and French fine-pitch jacquard cards   韦多尔及法式细针距提花机纹板
⑦ a piezo-element based zero-harness jacquard   一种压电基础零通丝提花机

based systems.

Miscellaneous carpet machines

Karl Mayer offers a carpet knitting machine for jacquard design loop-pile carpets which produces at a speed of 250 course per minute. The weft threads of the patterning pile are interlaced throueh the pillar stitches of the warp yarn. The four dead pile yarns lie in extended form in loop of the patterning pile. The weft thread covers carpet backing. This carpet with 5 or 8 pile tufts per inch has had success in the contract sector.

Cotinfi[①] offers a machine for producing carpets with adhesive-bonded cut or loop pile using an upgraded yarn folding technique (without holding blade). A new technique is the adhesive bonding of nonwoven webs[②] to produce loop-pile carpets, the consumption of PVC adhesives has been reduced from more than 1 kg/m^2 to 0.8 kg/m^2 with the incorporation of filler.

The adhesive is claimed to have greater penetration between the pile fibers and less penetration into the jute backing. The later can consequently be reduced from 407 kg/m^2 to 270 kg/m^2.

Vocabulary

tufting machine 簇绒机
linear 线性，线型
guillotine 剪切机
palette 调色板
stuffer warp （地毯的）衬垫经纱
back 织物背面
double lift 复动式
colorways 配色色位
flexible rapier 挠性剑杆
predominantly 主要地，显著地，流行地

piezo 电压
dust-tight 防尘的
absence of 没有
light guide 光控制，光波导
knitting machine 针织机
course 线圈横列，组织循环
non woven 非织造的，无纺织的
filler 填充料
sliding needle bar 滑动式针座
blade 刀片

---

① Gotifi （英国）Cotifi 公司
② the adhesive bonding of nonwoven webs 非织造布的黏合

cut pile 割绒
chain warp 链经
pattern definition 花纹清晰度
dead pile（地毯的）埋头绒头
open-shed 全开口
patterning 形成花纹，组成图案
scraper 刮刀，刮板
jute 黄麻

flexural 弯曲的，挠性的
encapsulate 密封
fine pitch 细针距
miscellaneous 其它
loop pile 起圈地毯
pillar stitch 编链经纱
adhesive 胶黏剂，黏合剂
penetration 渗透，穿透率

Think or answer these questions

1. What is the major feature of the "TX 300" tufting machine?
2. How to minimize cutting waste in weaving with the carpet weaving machine?
3. Which point of the carpet weaving machine should the dust extraction systems be put on?
4. Which advantages have the piezoelectric elements in jacquard engineering?
5. Who is the first company to offer a piezo-element based zero-harness jacquard?

UNIT 13 TESTING AND MEASURING EQUIPMENT
试验与测量仪器

LESSON 25
TESTING AND MEASURING EQUIPMENT(1)

Textile raw materials testing

Textile machinery makers have obviously been very concerned with detection of extraneous fibers in cotton fiber preparation and launched a number of systems. Zellweger, a prominent maker of testing apparatus, has also got to grips with the problem and offers the "Optiscan"① extraneous fiber detection system as a machine for spinning preparation.

---

① Optiscan （瑞士）Zellweger 公司的 Optiscan 异纤维检测系统

Contaminants in raw cotton such as fragments of polypropylene bale material, string, colored lumpy material, fragments of fabric plus the resultant defects in yarns, fabrics and finished goods have for a long time been the cause of costly complaints.

– **Fiber Contamination Tester**[①]

A minor sensation for testing equipment makers was created in 1995 in Milan[②] by the "Fiber Contamination Tester" launched by Maschinenfabrik[③] Rieter as an off-line tester for honeydew or stickiness in cotton. The device developed in Israel[④] by Dr. Uzi Mor[⑤] was in fact described back in 1994 in Bremen[⑥] in the course of a paper resented by its inventor.

Since there has as yet been no successfully operating device for measuring "honeydew" (NIR[⑦] measurement has so far not managed to gain acceptance) it can be justifiably claimed that this represents an innovation. It is also possible for the device to perform rapid tests for neps, trash and seed-coat fragments.

In recent times the problem of honeydew has affected more cotton origins than was the case several years ago. An efficiently operating reliable device for the early detection of stickiness as early as the ginning stage in the country of origin or at least before bale opening in the preparatory stages of spinning would be very helpful. But a problem that must not be overlooked is representative sampling with sporadically occurring honeydew.

① Fiber Contamination Tester 纤维杂质测试仪
② Milan （意大利）米兰
③ Maschinenfabrik （德语）机械制造厂
④ Israel 以色列
⑤ Dr. Uzi Mor 尤兹·莫尔博士
⑥ Bremen （德国港口城市）不来梅
⑦ NIR 近红外光谱测试

YARN AND FABRIC FORMING

– The testing of baled cotton

HVI ①cotton testing lines (High Volume Instruments) have gained worldwide acceptance for the testing of baled cotton. According to a current ITMF② report more than 1200 lines are operating throughout the world. At the time of the Hanover③ ITMA there were about 500 lines, of which almost half were in USA.

Powerful stimuli for the use of HVI lines have come not only from spinners worldwide or the on-line programming systems for quality optimization or blend composition, and from spinning machinery makers (in particular from Schlafhorst), but also from the cotton producing countries. In USA, HVI measurement forms the official basis for the grading of Upland cottons. In the long term other leading cotton producing countries are sure to follow.

Since the Spinlab④ and Motion Control/Peyer groups⑤ have now been taken over by Zell weger Uster, there is now only one maker of HVI lines in the world. The latest HVI line can be operated by one person and in a slightly modified version also offers the facility of quality assessment for man-made staple fibers.

Production control and optimization of the preparatory stages of spinning

An appropriate production control tool, especially for card optimization in spinning preparation, is the "AFIS"⑥ system for the quick non-subjective measurement of neppiness. A new item is the quick single-fiber maturity tester developed from original ideas of the inventor of the "AFIS" system.

The tried-and-tested "Ⅱ C-Shirley Fineness/Maturity Tester"⑦ has been further developed and was likewise exhibited in Milan in the form of the "Micromat" highspeed tester. The device with accelerated test cycle can be combined with existing HVI lines.

A device that has likewise proved successful for production optimization in rotor

---

① HVI （美国）动力控制公司研制的多功能快速纤维测试系统，主要用于棉花分级室，如 HVI3000 能在 25 秒内测定一份样品的细度、色泽、长度、长度不匀率、强力、伸长和杂质共七种指标
② ITMF （International Textile Manufacturers Federation 的缩写） 国际纺织制造者联盟
③ Hanover （德国）汉诺威
④ Spinlab （美国）思彬莱公司
⑤ Motion Control/Peyer groups 动力控制公司/佩耶公司集团
⑥ AFIS 高级纤维信息系统。一台 AFIS 棉纤维品质检测仪可以取代棉检室的全部工作，而且一人即可以操作
⑦ Ⅱ C-Shirley Fineness/Maturity Tester Ⅱ C-锡莱纤维细度/成熟度测试仪

spinning, especially raw materials selection, is the Ouikspin system developed by the Institut fur Textil-und Verfarenstechnik (ITV) Denkendorf. This is sold as the "MD-TA" system. Apart from dust, trash and fiber fragment measurement, predictions of spinning performance and expected yarn quality are possible. Blend optimization is possible, not only for three-roller spinning but even for woolen and semi-worsted spinning (at least up to 60 mm fiber length). Quality assessment of short-staple flax blend component is a further option, as demonstrated by trials in Reutlingen①.

Vocabulary

concemed with 参与，干涉
contaminate 杂质，污染物质
string 细绳，带子，线
resultant 总的
complaint 毛病，疾病，障碍
off-line 离线（工作线之外），脱机
stickiness 黏着性,附着性,胶(粘)性
trash 杂质
ginning 轧棉
representative sample 代表性取样
stimulus 刺激流，促进因素
composition 合成，结构，组成
Uplands cotton 陆地棉
slightly 稍微，有一点
neppiness 棉结
original 原始的,最初的
get to grips with 努力钻研，认真对待
prominent 重要的，著名的

fragment 碎片
lumpy 块状的，成块的，成团的
for a long time 长时间，长久
tester 测试仪
honeydew 虫污
as yet 现在还，到目前为止
in recent times 在近期，在近代
preparatory 准备的，预备的
sporadically 偶尔发生的，时有时无的，分散的
optimization 最优化，最佳参数选定
in particular 尤其，特别
grading 分级，分等
in the long term 从长远的观点来看
appropriate 适当的，合适的
maturity （棉纤维）成熟度
as flex 亚麻

---

① Reutlingen （德国）罗依特林根

Think or answer these questions

1. Can the "Fiber Contamination Tester" only used for measuring "honeydew"?
2. Is the "Fiber Contamintion Tester" an on-line tester?
3. Where the stimuli for the use of HVI lines come from?
4. How many persons to operate the latest HVI line?
5. What uses is the Quickspin system applied to?

LESSON 26
TESTING AND MEASURING EQUIPMENT(2)

A great deal of interest was attracted by a range of "low-cost" testing devices, especially from textile mill unwilling or unable to invest immediately in HVI lines. These are based on internationally successful predecessors by the Spinlab company (Fibrograph 730①, Colormeter 750②, Micronaire 775 ③and Fiberglow 380④). The last-named in particular should be useful for the assembly of color-sensitive cotton batches (for velvet!) and even for assessing additives on man-made fibers.

Man-made fibers can also be tested using a modified "AFIS" system. A very quick result of fineness and length distribution in the raw material is obtained with better statistical confidence than with conventional single-fiber measurement, since 2000 to 3000 fibers can be measured in just a few minutes.

For traditional single-fiber measurement of man-made fibers, still specified by the BISFA⑤ standard, a new instrument introduced was the "Vibroskop 400". In preparation for measurement the single fibers simply need to have clamps attached and be suspended in a frame, following which the test runs automatically.

---

① Fibrogroph 730　730 纤维长度照影仪
② Colorimeter750　750 色泽仪
③ Micronaire 775　775 马克隆尼气流式纤维细度测试仪
④ Fiberglow 380　380 纤维紫外线荧光测试仪，它可以快速、准确地测试棉和化学纤维的荧光，可作为控制纤维可染性能工艺程序和其他纤维混合的依据等
⑤ BISFA　国际人造纤维标准化局

On-line control of sliver ever evenness as a correction variable for drafter control has long since become an essential selling point for well-known makers of drafters. Further assistance is offered by control systems by leading machinery makers such as "Sliver master" (Loepfe)①, "Sliver Dance" (Barco) or "Sliver Contrl" (Zellweger) to name but a few. Centralized data collection offers opportunities here too for timely production and quality control. Some monitoring and evenness control systems were developed by the machinery makers, fitted directly on feed mechanisms of various opening and cleaning machines, with a growing trend towards ease-of-operation.

Yarn manufacture and yarn testing
– Spun yarns

Monitoring of the individual spinning head is frequently adapted on the rotor spinning machine in cases where yarns of perfect quality are required for direct downstream processing. The ring spinning frame still presents problems, especially in linking systems, although initial attempts at spinning head monitoring on the ring spinning frame have been known for some considerable time. Earlier solutions have mostly foundered on the obstacle of costs.

Maschinenfabrik Zinser in 1995 cooperated with Schlafhorst, Loepfe and Swiss spinner H. Bubler②in exhibiting a new spinning head identification system in which the separate cop holders are fitted with a freely programmable chip③, each combining quality and production data for the spindles in question plus quality data for the winder head. It also identifies defective spinning heads and sorts out defective cops according to pre-

---

① such as "Slivermaster"(Loepfe)... such as 后面为列出的系统名称,括号内为其制造公司
② Swiss spinner H. Bubler 瑞士纺纱工 H. Bubler
③ ... the separate cop holders are fitted with a freely programmable chip ……单独的管纱支架器安装有一个可任意编程的集成电路块

defined tolerance limits-①a further step in the direction of perfect yarn production.

One method shown in Milan that immediately caught the eye②, simulating woven and knitted fabrics and predicting possible defects in the finished fabric, wss the highly rated new "CYROS" system③ by Zweigle④in Reutlinge. This uses actual yarn data provided by "G580" optical yarn evenness tester. The CAD experience of the CIS⑤ company in woven fabric design and the experience of Cotton Inc⑥. in cotton and its processing brought about amazing reality in the simuiations displayed. The use of the same method, at the winder head or the rotor head, where optical sensor can be used too, is surely not far away.

– Filament yarns

A new generation of instruments was developed by Du Pont⑦ for testing filament yarns. Automatic yarn fineness testing, draw force measurement and interlacing testing, plus automatic shrinkage force measurement under the effects of heat on POY, MOY and LOY yarns and textured yarns were also offered by the traditional makers of filament yarn testing instruments.

These makers again concentrated on their strong points and introduced noteworthy improvements. For instance the "Dynafil M" apparatus for draw-force testing and shrinkage force measurement is equipped with automatic feed package changer (holding up to twenty packages). Packages are changed within two seconds with the aid of a splicer.

In texturing, thread tension monitoring at the texturing head is becoming increasingly adopted as a form of on-line quality control.

Successful quality data collection in the weaving plant will not be discussed in detail here, but the "Tensojet" breaking strength tester deserves a mention. This performs

---

① ... identifies defective spinning heads and solts out defective cops according to pre-defined tolerance limits ……
识别有毛病的纺纱头和根据预先规定的公差限制选出管纱
② immediately caught the eye   立刻吸引眼球
③ "CYROS" system   CYROS 织物模拟软件系统
④ Zweigle   （德国）茨威格尔公司
⑤ CIS   CIS 公司
⑥ Cotton Inc   （美国）棉花公司
⑦ Du Pont   （美国）杜邦公司

up to 30000 breaking tests per hour on yarns. thereby making possible the detection of random weak place in packages to a degree previously virtually impossible①.

Vocabulary

predecessor 以前的东西，前任
color sensitive 感色灵敏的
statistical 统计学的，统计的
suspend 悬(挂)，吊，挂
variable 变量，变数
long since 很久以前，早已
spining head 纺纱头
tolerance 公差，容限
optical 光学的
a range of 一系列，一排
last-named 最后提到的

additive 加成的，助剂，添加剂
confidence 可信度，可靠程度
evenness 均匀，均匀度
assistance 辅助设备
centralize 集中，中心
attempt 尝试，试图，试验
chip（集成）电路片，集成电路块，芯片
rated 评价的，设计的
interlacing 交缠
sort out 捡出，选出

Think or answer these questions

1. Which predecessors were made by the former Spinlab company?
2. What standard is specified by for traditional single-fibers measurement of manmade fibers?
3. May the new spinning head identification system be programmed?
4. What simulation can the new "CYROS" system do?
5. Which form of thread tension monitoring is adopted at the texturing head, on-line or off-line?

---

① ... making possible the detection of random weak places in packages to a degree previouely virtually impossible.
……在一定程度上，在卷装内随机低强部位的检测成为可能,这在以前是不可能的

YARN AND FABRIC FORMING

现代纺织英语

UNIT 14 FLAT KNITTING MACHINE 针织横机

LESSON 27
FLAT KNITTING MACHINE(1)

General development

With the different makers of flat knitting machinery it was noticeable that a number of developments launched as innovations at earlier ITMA exhibitions have now become state-of-the-art. Development in flat knitting machines is becoming increasing directed at rationalization and making knit manufacture more flexible. The call is for "quick-response", i. e. responding more rapidly to market demand and producing quickly ever smaller batch sizes with wide variety of design at a high quality standard①. In some cases several machines demonstrated the wide versatility of fully-fash ioned design. The majority of flat knitting machines are now fitted with fully-fashioning facilities as standard equipment.

These include a second fabric take-up motion underneath the needlebeds, sinkers, thread length control and on compact machines a takeoff comb for knit starts on empty needles. Length control was first launched by Shima Seiki②(DSCS③) at OTEMAS'④ 85 and it is now used also by Stoll⑤(STIXX⑥). Universal⑦(starting in mid-96)and Protti⑧ for keeping the size of the knitted panel constant. The principle is based on measurement and correction of the yarn length in a knit-

---

① ...respanding more rapidly to market demand and producing quality ever smalller batch size with wide variety of design at a high quality standard. 市场需要作出更快地反应,以高质量标准进行各种设计,越来越小的批量化快速生产
② Shima Seiki （日本）岛精公司
③ DSCS 数字式线圈控制系统
④ OTEMAS （日本）大阪国际纺织机械展览会
⑤ Stoll （德国）斯托尔公司
⑥ STIXX 线圈校准系统
⑦ Universal （德国）环球公司
⑧ Protti （意大利）Protti 公司

ting course.

The yarn passes over a measuring wheel which delivers a certain number of pulses per revolution. The on-board computer monitors the length variation from the target length in the course and keeps the length of the knitted panel constant where required by adiusting the knockover cam.

Wider use of auxiliary needlebeds

Wider use of auxiliary needlebed which are arranged above the needlebeds and equipped with transfer points provides clear evidence of the efforts flat knitting machinery makers to rationalize fully-fashioned knitting in particular and the knitting process in general. This technique was demonstrated in Milan by four machine makers. Shima seiki and APM① offer compact machines with two auxiliary needlebeds whilst Rimach② and Comet③ machines have one auxiliary needlebed.

In the case of Rimach and Comet the third needlebed is divided in such a way that when transfer takes place the two auxiliary needlebeds can simultaneously be shifted outwards or inwards. When narrowing, the right and left hand knitted edges in one carriage course can thus be simultaneously moved inwards. Apart from higher productivity there are additional patterning facilities as with double-knits stitches can be "parked" on the auxiliary needlebeds.

New machine developments
– Knitting of one-piece sweaters

The technique was demonstrated back at ITMA'91 in Hanover although in opened-out form ("Integral" knittiag)④. And in Milan in 1995 for the fist time the zero-make-up sweater⑤ was demonstrated. In the case of Shima Seiki the one-piece sweaters were produced on two newly launched type of machine, the four-needlebed "SWG. X" and

---

① APM （意大利）APM 公司
② Rimach （意大利）雷马其公司
③ Comet Comet 公司
④ ... although in opened-out form("Integral" knitting) 尽管以开发的形式（"集成"的编织）
⑤ the zero-make-up sweater 零缝制的针织套衫

the double-needlebed machine "SWG. V", and on the "SWG. X" it took 35 minutes to knit a sweater.

Fully-fashioned or three-dimensional knitting① is especially advantageous in the technical textiles sector too. Flat knitting technology offers not only the opportunity for zero-make-up production but also high flexibility and product versatility compared with other methods of fabric production. In items subject to heavy wear it is possible to orient the threads in the direction of the force flowlines. Universal in association with a car manufacture offered an interesting example from the-car sector, a one-piece knitted car seat cover (seat, armrests, hesdrest) knitted on an "MC-748" with Power-Pressjack②. This and other activities of makers of flat knit machinery reflect the growing importance of knits in the technical market.

Vocabulary

flat knitting machine 针织横机
in some case 有时候
sinker 沉降片，弯纱片
course 线圈横列
knit 针织，针织物
fully-fashioned knitting 全成形针织，收放针针织
empty needle 空针
pulse 脉冲
on-board 在仪表板上，在[操作]台上
knock-over cam 脱圈三角
transfer point 移圈针

narrowing 收针
sweater 运动衫，针织套衫
flowline 流线
as with 正如……情况一样
target 目标，指标
auxiliary needlebed 辅助针床(板，座)
in general 总之，一般，通常
transfer 移圈
one-piece 整体的，一片的，单片的
orient 定(取)向，排列方向，定(方)位
seat cover 座罩(套)
subject to 在……条件下，假定，根据

---

① Fully-fashioned or three-dimensional knitting 全成形或三维编织
② Power-Pressjack 机头随动沉降片式压脚

Think or answer these questions

1. How does the author understand the QSC?
2. What place are the auxiliary needlebeds arranged?
3. Who first launched the length control?
4. How many times does it take to knit a sweater on the "SWG. X"?
5. Which advantages does the flat-knit technology offer?

LESSON 28
FLAT KNITTING MACHINES(2)

The carriage-less flat-knitting machine

Amongst the innovations in recent years it has been the "TFX" carriage-less flat-knitting machine by Tsudakoma① that aroused the particular interest of the people. The machine offers a number of new facilities in respect of flexibility, productivity and reliability in manufacture. The essential feature of the machine is the individual-needle con-

trol with every knitting needle being driven by its own liner motor②. Every needle can be freely controlled independently. In one course it is possible to knit a maximum of thirty from a total of eighty possible stitch lengths. The combination of large and small stitches is possible, which surely offers a wide base for new design structures. The advantages are impressive, especially with respect to potential increases in machine productivity. With a needlebed width of 122 cm and a 15 cm clearing height, the "TFK" can be compared with an eight-system flat-knitting machine. The time waste occurring with the

---

① the "TFX" carriage-less flat knitting machine by Tsudakoma 津田驹 TFK 无三角滑座针织机横机
② ... the individual-needle control with every knitting needle being driven by its own liner motor ……单针控制每根织针由自身直线电机传动

YARN AND FABRIC FORMING

reciprocating movement of the carriage is totally eliminated①. At stitch transfer, several zones in one course can be transferred simultaneously. These machines also offer new technological facilities in reliability.

The overhead direct feed reduces the overall wrap ansle of the yarn②, permitting lower more uniform thread tension. In needle control, for maximum smoothness inknitting there is a choice of 250 cam variants. In this way it is possible to make objective allowance for the specific knitting requirements of different stitch constructions or materials. For instance it is eminently suitable for knitting low-stretch or delicate yarns. The final point to be mentioned is the easy maintenance of the machine. The axial needle drive avoids the shear forces③ that used to occur in the needle track, which should reduce mechanical wear. Apart form the positive aspects mentioned, many questions still remain to be answered regarding design versatility, knitting reliability and the work capacity of the machine. These can only be answered by industrial experience.

– Higher-productivity

A striking feature was the enormous efforts being made by machine makers to further expand current flat-knitting technology in the direction of higher productivity and more flexibility. In this context Stoll offered its "CSM 340.6" four-system compact machine and Universal its "MC-888", the first eight-system flat-knit machine. New ground was broken with the launch of the Universal MC-800 series. The chassis is compact and of ergonomic design④, the novel modular carriage control forms an integral part of the carriage and undertakes the control functions necessary for knitting⑤, whist the on-board computer is responsible for data transfer.

The advantages are a lighter workload for the on-board computer, simpler trailing cable configuration and greater reliability in data transfer. An interesting improvement of detail was introduced by Stoll with the optimized intarsia thread control by which means

---

① The time wastage occurring with the reciprocating movement of the carriage is totally eliminated　因滑架往复运动引起的时间浪费被完全消除

② The overhead direct feed reduces the overall wrap angle of yarn...　机顶式直接喂入减少了纱线的总包角

③ The axial needle drive avoids the shear forces...　轴向针驱动避免了剪切力

④ The chassis is compact and of ergnnomic design...　机架结构凑型符合人类工程学方面设计

⑤ the novel modular carriage control forms a integral part of the carriage and undertakes the control functions necessary for knitting　新型标准组件三角滑架控制机构,形成滑架的集成部分,承担编织需要的控制功能

the maximum speed has been increased from 0.7 to 1.0 m/s. Machine gauge has been increased since the ITMA'91 by several machine makers to E14, and in the case of Shima Seiki to E18.

The trend to greater flexibility and expansion of machine functions continues without interruption For instance the split Stitch Function (machine division) by Shima Seiki is now state-of-the-art. The same applies to resistance to laddering which has been dealt with in different ways by the leading machine makers in control respects. Networking of flat-knitting machines with the design studio①is now available from Stoll and also Universal. For knitting plants already having this facility it represents a vital tool for production monitoring and defect analysis.

Design systems

Just as with the machines, developments in design systems are focused on quick-response. programmes are becoming increasingly easy to use. Automatic modules facilitate programming and reduce the effort required in producing control programmes②. Great efforts have been made here especially in the field of fully-fashioned knitting. Simulation of stitch structure is now state-of-the-art and with certain machine makers it is available as an option. In computer system two routs are being followed. With PC- based ③ system the idea is primarily flexible low-cost computer assistance.

The systems frequently run with "Windows", enabling applications from the "Windows" programme range to be used. In the case of flat knitting machine makers having design systems based on work-stations, the productive capacity of the system has central priority. This is how the growing demands of increasingly more sophisticated programmes can be met.

Vocabulary

knitting needle 针织用针，织针　　　　clear height 退圈高度

---

① Networking of flat-knitting machine with the design studio... 针织横机与设计工作室的联网
② Automatic modules facilitate programming and reduce the effort required in producing control pro grammes 自动化组件促进了编程及减少了制作控制程序所需的努力
③ PC （Programmable Controller 的缩写）可编程控制器

stitch transfer 线圈转移
eminently 突出地，著名地
shear force 切力，剪（切）力
chassis 机架，机壳，机箱
split stitch 复合线圈，二重线圈
laddering（线圈）纵向脱圈
stitch length 线圈长度，线迹长度

reciprocating 往复（的，式）
wrap 包，缠
axial 轴（向）的
needle track 针道
intarsia 嵌花
machine gauge 机器隔距，机号

Think or answer these questions

1. What is the essential feature of the carriage-less flat-knitting machine?
2. Who is the company to develop the carriage-less flat-knitting machine?
3. How many systems is the M-888 flat-knitting machine offered by Universal?
4. Is net working of flat-knitting machines with the design studio now possible?
5. What does the design systems frequently run with?

UNIT 15 LARGE-DIAMETER CIRCULAR KNITTING MACHINES AND ANCILLARIES 大直径圆型针织机及辅助装置

LESSON 29
LARGE-DIAMETER CIRCULAR KNITTING MACHINE AND ANCILLARIES(1)

Greater flexibility in circular knitting machines

Today there is a growing trend towards circular knitting machines offering greater versatility in the manufacturing process whilst maintaining a high quality standard. High flexibility coupled with short change-over times and rational (economically viable) manufacture are especially advantageous in the production of small batch sizes both in piece-goods and fashioned panels. Flexibility for the machine maker means being able to use immediately the right machine for the product currently required by the market-place. Flexibility also means having the facility, even with a limited set of machinery, for converting machines to product-oriented production within a relatively brief time[①]. This is basically nothing new and in the past many diverse solutions have been proposed, but demands are becoming increasingly more stringent with short-lived fashions and search for new market outlets.

High flexibility in the manufacture of different products is summarized by machine makers by the "quick-change" concept. Knitters currently associate this concept mainly with:

— changing machine gauge (needle spacing),

— changing the diameter of the cylinder and dial,

— compatibility of modules between machine types,

— quicker changing of knit structures,

— quicker pattern changes (e.g. color-jacquard or striping selection) and

— high standard of user-friendliness and ease of maintenance.

— Conversion of gauge and diameter

The "quick-change" approach is already evident with a number of machinery makers in a modified machine frame (e.g. Orizio[②], Mayer & Cie[③], Monarch[④], Pilotelli[⑤], Terrot[⑥]). In the case of large-diameter circular knitting machines the frame is usually of the standard three-legged portal type which can be converted with little effort from 30 kg or 50 kg fabric rolls to 120 kg fabric rolls, plus some pedestal types in body width machines. Frame sizes are restricted to just a few within which a change of diameter can be made by using cylinders or dials of different dimensions.

---

① Flexibility also means having the facility, even with a limited set of machinery, for converting machines to product-oriented production within a relatively brief time 灵活性也意味着即使只有有一套设备,适于在相当短的时间内将机器转化成有关产品的生产

② Orizio （意大利）奥利就公司

③ Mayer&Cie Mayer&Cie 公司,主要产品为圆型针织机

④ Monarch Monarch 公司

⑤ Pilotelli Pilotelli 公司,小型国际圆机生产商

⑥ Terrot （德国）德乐针织机械公司

The distance between two power-carrying pillars or feed wheel carrier ring supports is sufficiently great to enable the dial and cylinder to be removed horizontally. In recent years "quick-change" has been demonstrated primarily in association with plain jersey machines not having individual needle selection facilities, in other words plain knitting machines. Changing cam segments for example change the structure from three-thread fleecy fabric to single fleecy or plain jersey. Converting the gauge within feed groups is performed without changing needle cam and feeders.

The time taken to change the gauge is a few hours on plain-jersey machines whilst 1.5 to 2 days are needed for double-jersey machines[1]. Gauge changing facilities represent about 20% to 25% of the machine cost price and diameter changing 30% to 40%.

– **Knitted constructions without single-needle selection systems**

Now the customer can choose between low-flexibility high-speed machines for plain-jersey, rib or interlock (closed needle tracks necessary) which are well established for the production of standard goods or goods for printing, and more flexible machines capable of covering a wide spectrum of knitted constructions. Some high-speed machines running at 2m/s peripheral speed were also used, unhappy by their well-known high noise level. Peripheral speeds around 1.5m/s have now become standard. Design-change facilities from interlock for example on machines with change cams is now state-of-the-art. These machines are eminently suitable for producing piecegoods such us print fabrics. On machines needing frequent pattern changes, machine refit time can be reduced using cams with external changing facilities.

Attention was attracted by an eight-cam machine incorporating three-way technique[2], the cam of which can be switched to another setting in the slow mode ("RDS"[3] drop cam system). Du Pont has developed a new rib-knit construction in-corporating an elastane thread in every course. The base thread (cotton) is worked in every feed. The elastane thread is plated alternately only with the rib dial needles and in the next feed only with the cylinder needles. A modified cam construction as offered by Terrot is required for this purpose. The technique produces a high-stretch knit.

---

[1] The time taken to change the gauge is a few hours on plain-jersey machine... 改变机号所花的时间在平针织物机器上没几个小时……

[2] three-way-technique 三路技术,即在一次喂入中不编织、集圈或编织

[3] RDS 旋转式压针三角

Jacquards with electronic individual needle selection

The knitter was faced with a virtually overwhelming variety of new or upgraded machines with electronically controlled individual needle selection. The choice lay between electronically controlled single needle selection with virtually unlimited repeat size in the distinctly more cost advantageous two-way technique or the more expensive three-way technique.①

Vocabulary

circular knitting machine 圆形针织机
change-over 改变,转变,更换
piecegoods 匹头,布匹
propose 提议,建议,打算
summarize 概括,概述,总结
compatibility 兼容性,互换性
portal 龙门架
pedestal 支架,轴架
pillar 支柱
feed wheel 喂给轮,输线轮
plain 平的,素的,平纹的
needle selection 选针
fleecy fabric 起绒织物
double jersey 双面针织物,双面乔赛
interlock 双罗纹组织
external 外部的,外置的,附带的
rib knit 罗纹组织,罗纹线圈
feed 喂纱,输线,成圈系统
dial needle 针盘针

overwhelming 压倒的,不可抵抗的
ancillary 辅助设备
rational 合理的,理性的
stringent 严格的,迫切的
short-lived 寿命短的
cylinder and dial 针筒和针盘
three-legged 三脚的
fabric roll 布卷
carrying 垫纱,喂纱,吃线
carrier ring 喂纱器圆环
horizontally 水平地
jersey 乔赛(针织物统称),平针织物
three thread 三线
rib 罗纹
refit 改装,重新装配
drop cam 压针三角
elastane alternatively 交替地,相间地,轮流地
cylinder needle 针筒针

---

① between electronically controlled single selection with virtually unlimited repeat size in the distinctly more expensive three-way technique 用户可在有明显成本优势的无限循环规模的由电子控制的单针选针及双路技术或更昂贵的三路技术之间作出选择

be faced with 面对着，面临着

Think or answer these questions

1. What does the flexibility mean?
2. What shape is the frame of large-diameter circular knitting machines?
3. How much percent of the machine cost price dose the gauge changing facilities represent?
4. Which kind of machines has the well-known high noise level?
5. What is the three-way technique?

LESSON 30
LARGE-DIAMETER CIRCULAR KNITTING MACHINES AND ANCILLARIES(2)

Noteworthy developments

Double-jersey machines by numerous makers had their design versatility extended with stripmg or transfer attachments[①].

One double-jersey machine was equipped with a sinker ring with hold-down/knock-over sinkers. This enables designs to be produced with longitudinal stripes using few needles knitting in the dial. As in plain-jersey machines this helps in restarting knitting following a press-off.

Single-jersey machines combined with 4,5 and 6-color striping attachments[②] were offered by numerous makers. Striping attachments with quick-fixing facilities promise ease in operation.

The range of plush machines suitable for producing outerwear fabrics and car and upholstery seat covers was extended with a much-admired color jacquard full-plush ma-

---

① ... had their design versatility extended with striping or transfer attachments ……设计的多功能性扩展了条纹或移圈装置

② single-jersey machines combined with 4,5 and 6-color striping attachments... 单面针织物机器与4、5、6色调线装置结合……

chine with which it is possible to produce large motifs with as many as twelve or a maximum of sixteen colors. Long floats and equally long pile are possible by using special sinkers.

Highly productive circular sweater machines with knitted welt were the direct competitors of flat-knitting machines. The machines(e. g. TLJ-6E by Jumberca①) with rotating needle track are showing the greater development potential②. Amongst the circular knitting machines with electronic individual needle selection they possess the highest standard of flexibility and the most versatile design potential through three-way technique in cylinder and dial stitch-transfer in both directions, five-color striping attachment, racking by plus or minus five needles and hold-down sinkers and latch openers for starting to knit on empty needles. Also of interest are devices such as in-and-out striping attachments for reducing yarn consumption when not using the full circumference of the machine.

Intarsia patterns could up to now only be produced on flat-knitting machines. Mayer&Cie developed an intarsia reversible large-diameter circular knitting machine③ capable of producing up to forty reciprocating movements or course per minutes with twenty feeds. With this machine it is possible to produce symmetrical intarsia motifs at a speed three to six times greater than would be possible on flat-knit machines.

Low-make-up underwear without side seam can now be produced on speciality plain-jersey machines with the device for knitting semi-fashioned panels with welt.

---

① Jumberca （西班牙）珍宝家公司
② The machines(e. g. TLJ-6E Jumberca) withrotating needle track are showing the greater development potential. 具有旋转针道的机器(例如珍宝家公司的 TLJ-6E 型)正显示出巨大的发展潜力
③ Mayer&Cie devdoped an intarsia reversible Large-diameter circular knitting machine... Mayer&Cie 公司研制了一种嵌花双面大直径圆型针织机……

YARN AND FABRIC FORMING

Wider patterning facilities in the knitting of fine-rib goods are available on machines capable of producing high quality fine-rib jacquard alternating with 2:2 rib without lowering the knit quality, transfer machines with and without welt or border were widely offered.

It is remarkable that not all machine makers have succeeded in incorporating electronic patterning in the machine legs. Sometimes large control modules impede access or view. Electronic systems incorporated in the machine frame represent an exemplary solution.

Electromagnetic needle selector systems by the well-known component supplier Harting① were used in numerous machines. Selection is made via one or two needle butts.

Design programmes of various machine makers can be compiled on CAD systems or also directly on the machine②. Few complete design systems are now offered by the machine maker but rather software for PCs.

Jacquards with mechanical individual needle selection

Many of the single-jersey and double-jersey machines are offered with mechanical jacquard needle selection as an alternative to fully electronic needle selection③, in which case the pattern width is generally 36 needles or 72 needles. Mini-jacquard selection systems in two way or three-way technique, and pattern wheel, druins, discs and combs are available from major and also many new smaller suppliers.

This technique has proved successful for a particular product spectrum such as textured constructions, minijac pattern etc④, it is simple and robust, has positive selection, is appreciably cheaper, it manages without any expensive infrastructure such as CAD ete. and sometimes permits higher peripheral speeds.

---

① Harting　Harting 公司

② Design progrnmmes of various machine makers can be complied on CAD system or also directly on the machine　不同机器制造商的设计程序能在 CAD 系统上编译，也能直接在机器上进行

③ ... with mechanical jacquard needle selection as an alternative to fully electronic needle selection　用机械提花选针作为全电子选针的替换物

④ for a particular product spectrum such as textured constructions, minijac pattern etc　适于特殊产品方面如花式组织结构、小型提花图案等

Vocabulary

transfer attachment 移圈装置
hold-down 握持,止动,压紧(装置)
press-off 脱套,脱圈
outerwear 外衣
long pile 长毛绒
intarsia pattern 嵌花花纹
reciprocating 往复运动
side seam 边缝,掖缝
patterning 形成花纹,组成图案
impede 阻碍,妨碍,阻止
compile 编制,编辑,编译(程序)
pattern wheel 提花轮
pattern disc 提花圆盘

sinker ring 沉降片圆环
longitudinal strip 纵向条纹
sinsle jersey 单面乔赛,单面针织物
motif 花纹图案,主题花纹
latch opener 开舌针
up to now 直到现在,到现在为止
symmetrical 对称的
fine-rib 细罗纹
Iransfer machine 移圈针织机
needle butt 针踵
mini-jacquard 小型提花
pattern drum 提花滚筒
comb 提花梳片

Think or answer these questions

1. Why was one double-jersey machine equipped with a sinker ring with hold-down/knokover sinkers?
2. Could intarsia patterns only be produced on flat-knitting machines?
3. Have all machine makers succeeded in incorporating electronic patterning in the machine legs?
4. Do the machine makers offer complete design systems?
5. What is as an alternative to full electronic needle selection?

LESSON 31
LARGE-DIAMETER CIRCULAR KITTING MACHINES AND ANCILLARIES(3)

Miscellaneous development trends

There was a marked development tendency to the machines of coarser gauge for outwear with the rustic look at present.

Programmable frequency-controlled three-phase drives① have now become state-of the-art and permit precisely definable smooth running and optimum speed at bothfull speed and at jogging speed②. A cable-less hand control for start/stop of circular knitting machines was offered③.

The sinker wheel has been moved by many makers into the outer feed zone which has facilitated conversion tasks.

Replacement of the fabric take-up roll by a cutting device with a container of 50 kg capacity can be completed within fifteen minutes.

Groz-Beckert developed a "meander" low-wall needle④, the recesses of which were filled with plastics. This prevents industrial dirt building up in the grooves without affecting the needle's already known benefits. Optimized knitting needles were also offered with conical hooks to minimize the risk of needle deformation in knitting.

Better rounding-off of edges on sinker and eye-needles plus a high standard of wear protection contribute to greater durability in knitting elements and a more consistent resultant fabric.

Various makers were prominent with their minimized surface-area smooth machine-frame, well-lit working area and easily accessible knitting zone.

There was greater use of machine simulations for illustrating a particular operational status or maintenance intervals, through to displaying the necessary gear-wheel pairings for fabric take-up.

---

① Programmable frequency-controlled three-phase drives 可编程的三相频控传动
② optimum speed at both full speed and at jogging speed 全速和慢行时的平稳运行和最佳速度
③ A cable-less hand control for start/stop of circular knitting machines was offered 一种用于圆型针织机起动/停止手持无绳控制器已推出
④ Groz-Becket developed a "meander" "low-wall" needle Groz-Beckert 公司研制了一种"曲折""低壁"针

Lint blowers, oilers, speed etc are monitored and regulated as required by programmable central control systems. Many suppliers are attempting to reduce machine costs by low-cost systems such as gear or v-belt driven fabric take-up. Almost all knitting machine makers offer their machines with an articulated fabric take-up and rolling device.

A trend that has intensified since ITMA'91 is the lowing of machine costs by standardization of the machine frames. ①

Ancillaries for circular knitting machines

Novel "hybrid" feeds were offered that can be used for either regular (positive feed) or intermittent (jacquard) yarn consumption.

New elastane feeds can now be loaded more easily with spools weighing as much as 1000g and have better protection from lint deposition.

The increasing use of creels having tube yarn guides adds interest to a tape welding device which permits feed wheel drive tapes to be welded together directly on the machine.

A further quality assurance opportunity is offered by continuous control of yarn run-in lengths. The prescribed run-in length is monitored constantly by a sensor roller and compared with tolerance limits②. If these are exceeded the machine is stopped.

Familiar blower systems (traveling lint blowers and compressed air jets etc.) have been complemented with "flutter" blowers and traveling fans which are self-cleaning③. A drawback of these systems continues to be that the lint disturbed is distributed back in the workroom. The task of trapping or removing lint in encapsulated creels or on the knit-

---

① ... lowing of machine costs by standardization of the machine frames ……依靠机架的标准化来降低机器的成本

② The prescribed run-in length is monitored constantly by a sensor roller and compared with tolerance limits 通过传感器罗拉和监测规定的喂入长度,并与允许的极限进行比较

③ ... "flutter" blowers and traveling fans which are self-cleaning ……"振动式"鼓风机和移动式风扇(其为自动清洁式的)

ting machine is one tackled mainly by Memminger-IRO, Frohlich, Piloteli, Shelton and LTG. In encapsulated creels the trend is to use blower systems operating mainly with internal air. On the knitting machine two different systems compete. Luwa provides a special air outlet above the plain-jersey or rib large-diameter circular knitting machine from which the air flow downwards in the form of a "cap" intended to deposit fibers and lint on the floor. Memminger-IRO, Monarch and Shelton opt for resolving the lint problem on plain-jersey machines with a suction hood located above the knitting cylinder[①]. If the forecast amounts of lint extraction can in fact be attained then a marked reduction in defect cost can be expected.

As far as quality monitoring on the knitting machine is concerned there were no particular advances. The familiar techniques have been updated and improved. It could become the accepted practice to supply a fabric scanner with every high-speed machine.

The productivity of circular knitting machines has attained a high level since ITMA'87. Today's opinion is that there is no need to increase performance by the number of feeds (up to five feeds per inch have been achieved) or the operating speed (up to 2m/s on large-diameter machines). Because of smaller batch sizes and quick-response to market needs the greater importance is now attached to higher flexibility. The compound needle represented a technical advance. But the extra cost is not compensated by the improvement gained over the latch needle, the performance of which has been greatly improved.

Vocabulary

replacement 取代,置换
building up 积累,堆积,结垢
eye-needle 眼子针,导纱针
oiler 加油器
intensity 加剧,强化,增强

feed wheel 喂给轮,输线轮
tolerance limit 容许限
compound needle 复合针
coarser-gauge 低机号,粗针距
sinker wheel 沉降轮,弯纱轮

---

① ... resolving the lint problem on plain-jersey machines with a suction hood located the knitting cylinder ……用一个位于在针织针筒上方的抽吸罩解决飞花问题

recess 凹座,凹进处
conical hook 锥形针钩
lint blower 飞花吹拭器,气流清洁器
articulated 铰链的,铰接的,有关节的

spool 有边筒子,线轴,筒子
prescribe 规定,指示,命令
trapping 截留,捕捉,收集
latch needle 舌针

Think or answer these questions

1. What use were the recesses of the "meander" low-wall needle filled with plastics?
2. What do the machine makers intend to lower the machine costs by?
3. What is the drawback of the familiar blower systems?
4. Were there particular advances as far as quality monitoring on the knitting machine?
5. Which side is the greater importance of the circular knitting machines now attached to?

UNIT 16 WARP-KNITTING AND RELATED MACHINERY
经编及相关机械

LESSON 32
WARP-KNITTING AND RELATED MACHINERY(1)

General assessment

Warp-knitting machinery comprises the following groups: automatic warp-knitting machines, raschel machines, stitch-knit machines, crochet galloon machines plus speciality and one-off modules①. The technical high-spots and the pattern of development are dominated by small number of market leaders. Suppliers in the "speciality" sector serve niche markets②. Cost and selling price constraints are reflected in the technical standard. Technical reliability and keeping close to the leading developments are prime considerations.

---

① one-off module 一次性组件
② Suppliers in the "speciality" sector serve niche markets 在"特殊性"方面的供应商为利基市场服务

YARN AND FABRIC FORMING

The constructional and design facilities provided with warp-knittechnology and the fabric engineering options are virtually unlimited①. Developments and innovations show that these advantages are understood by the warp-knitting machinery industry as the central challenge and opportunity. The overall aim is to enhance competitiveness:

— Maximum fabric production speed for piece goods without any design requirements in a wide variety of widths and excellent quality of appearance.

— Exploitation of constructional and design options by modifying knitting motions and their control, by varying the number and type of feeders, including weft insertion, and incorporating other textile(nonwoven webs) and non-textile materials in products.

— Innovative expansion of unique structural and design effects for the apparel, household and domestic textiles sectors and high quality textile structures for speciality technical end-uses.

— Non-damaging yarn conversion.

— Use of extremely different types and counts of yarns in one machine.

— Manufacture, with the need for making-up, of three-dimensional products, tubular items etc. using lapping techniques.

These technology and product related attributes are implemented with the aid of engineering developments(hardware) and extended software systems.

The emphasis is on economic benefits to the textile industry, namely high production flexibility, high production speed, quality assurance and reproducibility, quick design changes, user-friendliness, environmental compatibility (low-noise), electronic pattern processing and automated data collection.

The result achieved is the product of close cooperation with textile customers and end-users.

---

① The constructional and design facilities provided with warp-knit technology and the fabric engineering options are virtually unlimited 经编技术与织物工艺选择提供的结构和设计可能性,实际上是无限的

– Yarn feed systems

Depending on requirements, yarn feed systems exist in the form of computer-controlled positive mechanical feed mechanisms for high-precision reproducible yarn feed lengths and the sophisticated EBC① systems which provide sequential positive feed for producing patterning effects and controlling feed lengths. With the EBC system supplementary effects can be obtained by selective variation of thread tension/stitch length. Typical new and updated EBC systems are those incorporating servomechanisms with higher operational reliability in the event of mains supply fluctuations and power-failure.

Control of lapping systems for inserting weft threads via programmable servomechanisms② provides high quality weft insertion with little stress on the weft threads. Insertion rates exceeding 5000m/min do not represent the limit. Some solutions are aimed at minimizing edge-trim waste by making full use of the length of weft yarn inserted③ whilst some devices are intended for especially difficult weft yarns (e.g. glass filament yarn) with positive lapping of the threads by the specified increment.

With yarn-insertion systems too, carded webs and suchlike can be fed with electronic control and regulation. These yarn feed systems synchronized with an electronic guide bar control system and the cloth take-up motion, meet the most stringent requirements in respect of reproducible quality, but must be compatible with the textile product. Stitch-knit machine and to some degree also speciality warp-knit machines have caught up with state-of-the-art technology as far as feed systems are concerned④.

– Knitting elements and their motions

The emphasis of development towards higher productivity in the long term is still on knitting elements and their motions. By using modern materials, advances in light-weight

---

① EBC　卡尔·迈耶公司制的可程序控制的电子送经机构
② Control of lapping systems for inserting weft threads via programmable servomechanisms...　通过可编程伺服机构衬入纬纱的垫纱系统的控制
③ minimizing edge-trim waste by making full use of the length of weft yarn inserted　充分利用衬入的纬纱长度来减少剪边的浪费
④ Stitch-knit machines and to some degree also speciality warp-knit machines have caught up with state-of-the-art technology as far as feed systems are concerned　针编机在一定程度上也是特殊性的经编机，就喂纱系统而言，已经赶上了现代技术

construction, FEM-computed bar profiles① and computerized optimization of knitting motions, further increases have been successfully achieved here (>3000 courses/min. on high-performance automatic warp knitting machines).

Drive via crank mechanism ② is state-of-the-art for all high-speed automatic warp-knitting machines. It is noteworthy too that in raschel machines that used to be fitted with can mechanisms, there is a switch to crank mechanisms in the interests of low energy consumption, reduced heat emission, less noise, less stress on the yarn and final cloth quality.

Vocabulary

warp-knitting machine 经编机
raschel machine 拉舍尔经编机
high-spot 重点，突出部分
exploitation 开发（采，拓），发掘
domestic 家庭的，国产的，自造的
tubular 圆形的，管状的
power failure 电源故障
increment 增量，增加
guide bar 经编机梳栉
crank 针尾的弯曲段

in the interest of 为了，为了……的利益
warp knitting 经编
stitch knit machine 缝编机
constraint 抑（限）制，制约，强制
apparel 服装
conversion 转变，改变，转化
non-damaging 非破坏性的
edge-trim 剪边，修边
bar （经编机的）梳栉
heat emission 热辐射，热量发射

Think or answer these questions

1. Which groups does the warp-knitting machinery comprise?
2. The emphasis is on economic benefits to the textile industry, what does it mean?
3. What system is the EBC?
4. What is the emphasis of development towards higher productivity in warp knitting machines?

---

① FEM-computed bar profiles　　FEM 法(力学)计算的梳栉轮廓
② Drive via crank mechanism...　　通过针尾弯曲段机构传动……

5. As what use does the switch to crank mechanisms of raschel machines act?

LESSON 33
WARP-KNITTING AND RELATED MACHINERY(2)

The guide-bar mechanism can be driven by computer-controlled liner motors. In combination with the EBC system, lapping repeats are virtually unlimited with quick changes of pattern and sample production. The racking of the guide bars can be more finely graduated by computer control than by mechanical means①. In terms of drive technology, the future standard will be set by individual drive via toothed belts for lower noise level plus continuously variable drive control.

To increase product and patterning flexibility as well as user-friendliness, there are very many proposed solutions which provide easier adjustment of the knitting elements (e.g. conversion from pile or jacquard patterning to plain fabric, combined height adjustment of trace comb and knock-over comb bars on raschel machines etc.).

– Production data collection systems

Production data collection systems for warp knitting machines meet the state-of-the-art in this sector generally. Their purpose is to optimize efficiency (machine efficiency and quality in knitting plant) and materials management. Their utility value is unquestioned. The extension of electronic control to all groups of warp knitting machines involves a marked advance in the development of CAD② systems. The frequently far more complex warp-knit structures (e.g. open-work effects created by thread tensions③) and technical control requirements of adequately converting draft designs have for a long time caused warp knitting to lag behind weaving and weft knitting in this area. This problem has now been largely resolved. Realistic simulations on the monitor and natural printout colors are being developed④.

---

① The racking of the guide bars can be more finely graduated by computer than by mechanical means   用计算机控制比用机械手段更能精确地给经编机梳栉的分级
② CAD （computer aided design 的缩写）计算机辅助设计
③ open-work effects created by thread tensions   由丝线张力创造的起孔效应
④ realistic simulations on the monitor and natural printout colors are being developed   在监视器上逼真地模拟与打印出自然色彩正在开发

– Textile fabric constructions and design

The options for modifying the fabric structure and design in a warp-knit are just numerous as in any other technology. In this respect stitch-knit machines and speciality machines deserve special mention because of their speciality product facilities in the household textiles and technical textiles sectors.

Marked advances in technology with extended and sometimes completely innovative facilities for product construction and design for the textile industry are demonstrated by the following examples.

– Automatic warp-knitting machines with guide-bar control via liner motor for high-speed response. Important for low-cost production in high fashion sector for short lengths of elastic cord fabrics with pleated variants with open and closed fabric ground.

– Developments in the jacquard and multi-bar raschel sector wih new drive systems, for example.

– Single-thread control via piezo jacquard segments (no harness cords or magnets)[1], for plain or structured fabrics and embroidery effects knit-brocade without any cutting of threads.

– Patterning bars controlled by stepping motor.

– In the interests of high production speeds, low noise emission and maximum design flexibility.

– Developments for a very wide variety of technical textile production (e. g. biaxial semi-manufactures) using geometrically idealized yarn feed with tension control in the inlaid weft system too. Cloth take-up can optionally operate also in the raschel machine version with the opportunity for interesting product variants interlinings, leisurewear fabric etc.

– Stitch-knit machines with course-compatible weft reinforcement for high-speed lightweight stitch-knit production.

---

① Single-thread control via piezo jacquard segments (no harness cords or magnets)　通过压电提花部分(无通丝或磁力)控制单根丝线

— A jacquard carpet knitting machine for 5-frame boucle designs with non-patterning pile yarns fed into the pattern pile, outstanding clarity of appearance and wear resistance. A really valuable additional option to needled and tufted designs.

— An electronically controlled raschel machine for extremely heavyweight technical structures such as nettings, textile solar collector panels, geotexiles.

A simulation dislplay for monitoring different machine functions deserves special mention. This represents an initial stage aimed at remote diagnosis with the aid of multi-media technology①.

Summing up, it is estimated that thanks to their immense technical potential, warp-knitting and related machines have good chances of growth in the competitive economy.

Besides established end-uses, extension of technical applications will be governed to an increasing extent by developments in product specification profiles.

The use of high-tech yarns is progressively becoming more important here too.

Vocabulary

weft knitting 纬编
realistic 实际的，现实的，逼真的
pleated 褶间的，打褶的
multi-bar 多梳栉
stepping motor 步进电机
semi-manufactures 半成品
geometrically 几何学上，用几何学
interlining 衬头，中间衬料
wear resistance 耐磨性，耐服用性
boucle 珍皮呢（仿羔皮呢），结子线织物
solar 太阳的，日光的
estimate 估（计，算，量，价），预算

lag behind (in) 赶不上
resolve 解决，解答，消除
deserve 值得，应得，应受
elastic cord 松紧线，松紧绳
brocade 锦缎，花缎
biaxial 二轴的，双轴的
idealized 理想化的，形成理想的
inlaid 嵌花
leisurewear 家常服装，假日服装
clarity 透明度，清晰度
netting 结网，（船舶用）绳网
geotextile 土工纺织品，土工布
multi-media 多媒体

---

① remote diagnosis with the aid of multi-media technology 以多媒体技术为辅助的远程诊断

specification 规格，规程，说明书　　　　（be）governed by 取决于，由……决定

Think or answer these questions

1. Which method can the racking of the bars of warp-knitting machines be more finely graduated by?
2. Do the production data collection systems for warp knitting machines meet the state-of-the-art?
3. Are the options for modifying the fabric structure and design in a warp knit few?
4. Is the raschel machine for extremely heavyweight technical structures an electronicauy controlled one?
5. What does a simulation display for monitoring different machine functions repre-sent?

UNIT 17 EMBROIDERY 刺绣

LESSON 34
EMBROIDERY（1）

Although the boundaries between Schiffli and single and multi-head embroidery machines are gradually becoming less distinct as far as end-uses are concerned. the two systems will be considered separately here.

Schiffli machines

Two trends are evident in Schiffli systems. The first could be headed:"The allround embroidery machine is dead, long live specialization"[①]. In fact all Schiffli embroidery machines to date are designed in such a way that any of them could be capable of producing the entire gamut of embroidery products, from open-work, net, pierced and allover designs, through handkerchiefs, lace covers, curtain edging ("macramé lace) and curtai-

---

① The all-round embroidery machine is dead, long live specialization　万能刺绣机已经不适用了，取而代之的是专门化设备。

ning, to multi-color products such as tapestry, logos and badges of all kinds. With a good basic machine set-up only a few manual adjustments are needed to change the tensions of face-thread and under-thread.

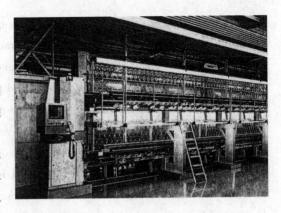

The trend now developing, however, is aimed towards supplying machines for a limited end-use and equipping these for optimum produetivity otherwise unattainable. Perfect examples of these are the Lasser L77-115[①] and the Saurer"Epoca"[②]. The L77-115 has an embroidery length of 30.8 yards = 28.5 metels and is thus the longest embroidery machine ever built. Deducting the unused unembroidered materials for lateral hooking-up, the length of 27 meters is precisely that demanded by customers buying curtaining.

The advantages are obvious. There is no need to interlace and frame-mounting is reduced to one operation lather than three(with ten-yard material) or two(with fifteen-yard material). But what is far more important, joins are unnecessary, for despite taking the utmost care join marks creep in where new embroidery adjoins existing embroidery[③], faults which can seldom be remedied and thus cause downgrading. A further factor not to be underestimated is the appreciably lower skill level required in operatives.

The drawback of this mammoth machine are that it is only suitable for repeats of 12/4 and greater, the maximum speed is comparatively modest 155 min^{-1} and the embroidery frame height is"only" 115 cm so that curtaining can only be embroidered to a height of about 110 cm without reframing. Purchase of a machine such as this invariably means the construction or acquisition of a new building for the purpose.

---

① Lasser L77-115   Lasser L77-115 型刺绣机

② Saurer"Epoca"   绍雷尔公司的 Epoea 高性能有梭绣花机,它采用了"基础的新技术",供中到小尺寸的批料使用

③ for despite taking the utmost care join marks creep in where new embroidery adjoins exiting embroidery   因为尽管加最大的小心,在新刺绣与现存刺绣连接的地方会产生连接痕迹

YARN AND FABRIC FORMING

现代纺织英语

The Saurer "Epoca" on the other hand sets different priorities. Using "Actifeed"①, an entirely new design of thread feed system, embroidery speed up to 400 min^{-1} are attained. The precise length of the next stitch is calculated by the control computer and the feed roller (segment of the earlier emery-covered roller) rotated by the appropriate increment②. It is thus possible to prescribe the angle of rotation compatible with the face material in question. Together with a completely new thread control system (there is only one thread guide) it ensures that tension peaks occurring in the face thread during the stitch-forming process are greatly reduced, so it is always possible to operate with optimum thread tension whilst at the same time reducing the incidence of thread breakage, despite the fantastic embroidery speed. Another contributor to this is the entirely new embroidery frame which has been replaced by a "frameless fabric support system"③.

The "Xtraline" support system for the two fabric rollers in a ten-yard machine has five-point drive from above and two point drive from the side (both on the front on the "Automaten"④) using servomotors. Another new feature is that rerolling of the embroidered fabric no longer has to rely on the sensitivity of touch of the operative⑤, it is controlled by the "QuickRoll" system⑥. Specially developed sensors monitor and ensure constant tension in the material. It is a great advantage that the machine no longer requires a conventional foundation, in other words it does not have to be "cast" in position⑦.

This new development is only in single-tier format. It is most effectively employed in embroidery plants using a large number of colors or materials, with the "PentaCut" thread trimmer device⑧ well-known from the 3040 module used to good advantage, for orders of

---

① Actifeed 一种积极式丝线喂入系统
② The precise length of next stitch is calculated by the control computer and the feed roller (segment of the earlier emery-covered roller) rotated by appropriate increment 下一个针绣的精确长度由计算机控制,喂入罗拉(较早期的金刚砂包覆的罗拉部分)按合适的增量转动
③ the "Xtraline" support system Xtraine 支架系统
④ Automaten (德语)自动化装置
⑤ Another new feature is that rerolling of the embroidered fabric no longer has to be rely on the sensitivity of the operative 另一个新特点是已刺绣织物的再卷不再依靠操作手感的敏感性
⑥ "QuickRoll" system QuickRoll 系统,在翻转过程中靠位置传感器监测织物的位置,织物张力是与产品相关可调的并可重复
⑦ ... the machine no longer requires a conventional foundation in other words it does not have to be "cast" in position ……机器不再需要传统的基础,换言之不必在适当位置"铸造"
⑧ "PentaCut" thread trimmer device PentaCut 线头修剪装置

medium to high volume. High-fashion embroidery is an important area. Conversely it is rather unsuitable and too inefficient for single-color relatively coarse embroidery such as "macramé lace" (curtain edging), and net or all-over curtaining for subsequent plain dyeing.

– **Electronic creel control**

Trend number two involves the continuing perfecting of the familiar tried-and-tested standard machines. Electronic creel control with increasing sophistication of details are taken for granted nowadays with all machine makers. With the exception of Comerio, repeat and color changing is now standard equipment with suppliers, likewise thread trimmer devices. Further improvements include enlargement of the embroidery field with greater frame height and wider traversing of the creel. A further example comes from Lasser, similarly to the non-operating needles the shuttles too are stopped, thereby preventing fraying and soiling of the bobbin threads that arte not needed.

Single-and multi-head automatic embroidery machines

The same trends are evident here as in Schiffli machines. On the one hand sensational new developments, on the other improvements and perfecting of industrially well established machines.

Vocabulary

embroidery 刺绣,绣花,刺绣品
Schiffli machine 席弗里刺绣机,飞梭刺绣机
pierced 纱罗
handkerchief 手帕
curtaining 彩色提花窗帘布
logo 边印
face thread 面线
deduct 扣(减)除,减(扣,除)去
frame mounting (在)框架上固定(安置)
adjoin (联,连)接,(附)加

fault 疵点,故障,错误
underestimate 低估,看轻
comparatively 比较地,比较上,稍微
embroidery frame 绣花绷,绣花架
priority 优先(权),重点,优先考虑的事
emery 金刚砂
fantastic 奇异的
inefficient 效率低(差)的,不胜任的
nowadays 现今,如今,现在
fraying 纱线滑溜,磨损,散边
boundary 边界,界面

gamut 全部,全程,全范围
allover design 满地花纹图案,满地印花图案
lace cover 花边罩
tapestry 绒绣,象景织物,提花装饰用毯
badge 徽章
unattainable 达(做)不到,不能完成的
despite 不管,任凭,尽管
creep 蠕变,塑性变形,(材料)潜伸
downgrade 使……降级,降低……等级
remedy 补救,修理(补,缮),校(纠)正

mammoth 巨大的,庞大的
modest 适度的,有节制的
reframing 再拉幅
acquisition 获(取)得
stitch 针绣
thread guide 导纱器,导纱钩
in position 在适当的位置
conversely 相反地,逆,倒
thereby 因此,所以,从而
bobbin thread 底线,梭花线

Think or answer these questions

1. Which embroidery products could the Schiffli embroidery machine be capable of producing?
2. How long is the precise embroidery length embroidered by the longest embroidery machine?
3. Does the embroidery machine require a conventional foundation?
4. Is repeat and color changing now standard equipment with suppliers?
5. What are the development trends of single-and multi-head automatic embroidery machines?

LESSON 35
EMBROIDERY(2)

– Multi-head machine with laser beams

Tajima ① claimed the most spectacular innovation with its multi-head machine e-

---

① Tajima Tajima 公司

quipped with laser beams①. A laser is fitted at the side of every embroidery head, by means of which embroidered portions of the design such as lettering that have just been embroidered outside the embroidered demarcation line can be trimmed off ②. It is thus possible for example to place first green and then red material on a white ground. First any figure is embroidered. The machine automatically shifts the material sideways under the laser and actuates the laser. It burns off the topmost, in other words the red, material at the desired distance from the contour line. It then moves a little outwards and burns off the green material. Creating a kind of shadow effect on the red material ③. The white ground remains totally unaffected.

The great advantage of these "TLFD" machines is that any material of any thickness can be trimmed, even different types within one design, without fraying of the edges. Simple "palette-knife" effects ④ are also possible of course. However, the cost of laser irradiation is still exceptionally hish so in the foreseeable future there is no prospect of machines with twenty or more heads. Prospects are more realistic for machines with a maximum of 8-heads.

The second new development to create a stir in⑤ Milan was the X25 240D multi-head automatic embroidery machine by ZSK ⑥ with portal and border frame and 25 heads. It was specifically designed for the embroidery of curtaining, having an embroidery length of about 6 meter and an embroidery field depth of max. 1000 mm. Framing naturally takes less time than on the Schiffli machine, but problems persist in joining-up following interlacing. "Florentine" net curtains⑦, in other words curtains with continuous patterning, are therefore unsuitable. But according to those interested in this machine there are some very definite proposals for using it in other directions.

---

① Multi-head machine with laser beams   带激光束的多头机器
② such as lettering that have just been embroidered outside the embroidered demarcation line can be trimmed off   比如绣在边界线外部的文字正好能被修掉
③ Creating a kind of shadow effect on the red material   在红原料上创造一种阴影效果
④ simple "palette-knife" effects   简单的"调色板刀"效果
⑤ create a stir in   在……引起轰动
⑥ ZSK（德国）   ZSK 公司
⑦ "Florentine" net curtains   弗洛伦廷网状窗帘

– Further upgrading and perfecting

As was the case with the Schiffli machines, in the single-and multi-head automatics there was further upgrading and perfecting of well tried-and-tested machines. One example was the multi-head automatic machine based on the tambouring machine with which it is possible to embroider chain stitch and moss stitch at an appreciably higher rote(up to 600 min^{-1}), which can be converted to chain stitch thereby creating "cord" stitch① and enabling narrow tapes and such-like to be sewn on. Some machine makers add a trambouring head to every standard head operating on the twothread systems.

These machines are then truly universal but whether it is more sensible for embroiderers to use two different machines will be revealed by industrial experience. Pfaff② claims to be "world record holder" as its machines are the first and only ones to operate continuously at 1000 stitch mi^{-1}. Almost all makers equip their machines with an electronic stitch-size dependent speed control system③, one of which is the "skip-stitch" technique④. With very long stitches the embroidery frame is just shifted laterally by half without the needle penetrating in this "interim stop".

This is far more effective than lowering the embroidery speed excessively which would otherwise be necessary for a speed easy on the machine and guaranteeing an optimum result. The "free-arm" models must not be overlooked in this context. Their purpose, following a fashion coming over to Europe from America⑤, is to embroider the front portion of peaked leisure caps or made-up T-shirts and sweatshirts, hand towels, entire bathroom sets etc.

---

① creating "cord" stitch 创造出"凸纹"线迹
② Pfaff Pfaff 公司
③ Almost all makers equip their machines with an elctronic stitch-size dependent speed control system 几乎所有的制造商都用一个针迹大小速度控制系统装备它们的机器
④ "skip-stitch" technique "跳针"技术
⑤ following a fashion coming over to Europe from America 仿效从美洲传到欧洲的一种时尚

The items to be embroidered are set-up in special frames. Whilst some makers equip their machines specifically for this one purpose, others convert "standard" multi-head automatic machines appropriately by lowering the tabletop. One more trend must finally be mentioned, namely the fact that today the embroidery of motifs and other designs is increasingly becoming a combination of printing and embroidery①. For example the images printed on T-shirts are simply then "enhanced" by a few discrete portions of embroidery. Naturally this demands sophisticated techniques for registering the embroidery at precisely the right location on each head. Besides there are many smaller companies supplying what are frequently essential ancillaries. Examples of these are the Applicut device for "pinking" off excess material in appliqué items, the Heinzle high-speed repeating machine with photoelectric reading, plus the core-less under thread spools and a novel patterned ground fabric which saves a large numher of cover stitches in emblems etc.

Vocabulary

laser beam 激光束
portion 部分，区段
demarcation 分[边]（线）界，划界线
burn off 烧去，烫去
irradiation 照射，辐射，热线放射
persist in 坚持，持续，持久
tambour 在绷圈上做（刺绣），在绷圈进行
chain stitch 链式针迹，链式线迹
reveal 展示，显示，揭示
peaked cap 无舌尖顶帽
T-shirt 短袖圆领衫，T恤衫
Table top 桌面
discrete 不连续的，分散的

appliqué 贴花，嵌花，加缝刺绣
core-less 空心
spectacular 惊人的，引人注意的，壮观的
lettering 文字
topmost 最高的，最上（面）的
contour 外形，轮廓
foresee 预见、预知
as in the case 通常那样
embroider 绣，刺绣
moss stitch 桂花针法
laterally 在[由]侧面地，横向地
leisure 丝绒边，绸缎边
sweatshirt 汗衫
convert 改造，改装

---

① ... is increasingly becoming a combination of printing and embroidery ……正在日益变成印花与刺绣的结合

pinking 锯齿切裁，衣边剪花
photoelectric 光电的
emblem（用作识别的记号或图案）胸袋
（上的）刺绣

Think or answer these questions

1. How does the multi-head machine with laser beams create a shadow effect?
2. Is there the prospect of machines with twenty or more heads with laser in the foreseeable future?
3. What embroidery product was the X25 240D multi-head automatic embroidery machine designed for?
4. What speed does the machine by Pfaff run at?
5. What is one more trend in embroidery sector?

UNIT 18 BRAIDING AND BOBBIN LACE MACHINES 编带及梭结花边机

LESSON 36
BRAIDING AND BOBBIN LACE MACHINES(1) FLAT KNITTING MACHINE(1)

In the braiding machine sector it was very much influenced by ISO 11-111[1] in respect of safety, dust emission and noise reduction in braiding machinery, although arguments are still continuing on what is actually required of machines, including cost aspect too. The effort of machinery makers continued to be directed towards increasing productivity by automation, higher bobbin capacity, the use of new materials and more powerful process monitoring.

Braiding machines from Cobra[2] are designed for markets ranging from very fine braids to the smaller ropes. The Cobra 250 is now claimed by the maker to run at higher

---

[1] ISO 11-111 （国际标准）纺织机械安全性要求 ISO 11-111
[2] Cobra Cobra 系列编带机

speed whilst the Cobra 450 with a new gentle-acting lacer motion is capable of maintaining braid quality whilst using yarns of inferior quality. Incorporation of combination of new structural materials and the mew lacer of the Cobra 250-80 enables very fine delicate yarns to be used①. The relatively wide spindle pitch of 80 mm permits high bobbin capacity with fine filament yarns. The lacer of the Cobra 540 has a higher bobbin capacity with wider bobbins whilst bobbin height remains unchanged. Nowadays the noteworthy feature of the braiding machines included encapsulation and a high standard of operational reliability.

The braiding machines of ETK Lesmo② incorporate exemplary safety and noise-reducing features. All moving parts including the unwinding motions are fully encapsulated for noise reduction and in the interests of comprehensive safety.

Herzon③ offered production machines ranging from thin braided cords to climbing ropes. The "SE 1/16-180" new generation of rope braiding machines now has a higher bobbin capacity of 2208 cm^3 (formerly 1400 cm^3). A new drive concept provides electronic control of the rotational speed of the flyer disc, yarn withdrawal, and wind-on with facilities for making changes whilst the machine is running. The largest machine offered is also designed for the production of climbing ropes, having a spindle pitch of 336 mm and equipped with 16 lacers, each with a bobbin capacity of 13271 cm^3. The length of lay can be kept constant with full bobbin as well as those almost exhausted using an electronic control system ④. The patented braiding eye indicates⑤ when a core yarn has run out, if the unwinding motion malfunctions, skewing due to irregular yarn tension or knots formed in the braid. An optical sensor stops the braiding machine when the lacer bobbin becomes empty. For monitoring productivity the machine can be equipped with a production information recording system.

---

① Incorporation of combinations of new structural materials and the new lacer of the Cobra 250-80 enables very fine delicate yarn to be used 新结构材料与 Cobra 250-80 新花边筒子组合的结合,能使用非常细弱的纱线

② ETK Lesmo ETK Lesmo 公司

③ Herzog Herzog 公司

④ The length of lay can be kept ooustant with full bobbins. as well as those almost exhausted using all electronic control system 捻距对整个筒子能保持不变,而这几乎排除了使用电子控制系统

⑤ The pectented braiding eye indicates... 专利的编织眼显示……

YARN AND FABRIC FORMING

Melitrex [1] developed its new twin-head braider with 130 mm spindle pitch (flyer disc speed 280-320 rpm). The bobbin capacity is about 800 cm^3. The machine is equipped with noise reduction facilities.

The O. M. A. braiders launched in 1995 [2] with 240 mm spindle pitch and 16 lacer, similarly to the newly developed twin-head "104 HCM/2" machine, are now equipped with electrical auto-synchronization between yarn withdrawal motor and flyer disc drive. These enable the braiding constant and the rotation of the flyer disc to be adjusted, including with the machine running. The yarn withdrawal motion and flyer discs can be advanced or reversed separately in slow motion for correcting braid defects [3]. The control console displays the rotational speed of the flyer disc, the production rate and the lay angle of the braid. The encapsulation and noise reduction on the sample machine were very good with a claimed 75 dB(A) noise emission. For the 240-pitch machine a new lacer has been developed with a bobbin capacity of 3200 cm^3, and a new thread tensioning system with a number of capstan rollers is located above the bobbin to maintain greater yarn compensation capacity [4].

Vocabulary

braiding 编织,编带,编结
bobbin lace 梭结花边
rope 索,绳
inferior 低级的,劣质的
braided cord 编织的细绳
flyer disc 锭翼盘
run out 用光,期满

skewing 斜,歪扭,偏移
autosynchronisation 自动同步
console display 控制显示器
braiding machine 编带机,编绳机,花边编织机
braid 滚带,编带,辫
lacer 花边筒子

---

① Melitrex   Melitrex 公司
② The O. M. A. braiders launched in 1995...   1995 年推出的 O. M. A. 编带机
③ The yarn withdrawal motion and flyer disc can be advanced or reversed separately in slow motion for correcting braid effects   纱线退绕机构和翼锭圆盘能以缓慢运动的方式分别前进与倒退适合于纠正编织疵点
④ a new thread tensioning system with a numher of capstan rollers is located above the bobbin to maintain greater yarn compensation capacity   一种新的有一些绞盘的丝线张力系统,位于筒子上方以维持更大的纱线补偿能力

exemplary 典型的,模范的　　　　braider 编带机,编结机
climbing rope 登山绳　　　　　　console 控制台,仪表板,托架
withdrawal 退绕　　　　　　　　capstan 绞盘,主动轮,曳引机
malfunction 不正常,故障,失灵

Think or answer these questions

1. What are the noteworthy features of the braiding machine today?
2. What uses do the braiding eye have?
3. What aims are all moving parts of the braiding machines of ETK Lesmo including unwinding motion fully encapsulated for?
4. What use is an optical sensor of the braiding machine for?
5. What use is a new thread tensioning system with a number of capstan rollers for?

LESSON 37
BRAIDING MACHINES(2)

The new generation of Retera machine was offered in modified design. A newly developed braiding eye monitor knocks off the machine in the event of yarn withdrawal failure or skew duo to irregular yarn tension. The lacers for the machines with 80 mm or 104 mm spindle pitch have higher bobbin capacity. Nylon inserts in the grooves of the flyer discs and rotating bushes between the two lacer discs ① are claimed by the maker to ensure smooth lacer motions transmission and reduce noise emission, hardened lacer feet and flyer discs reduce wear.

A first-time product was a triple-head cable braiding machine accepting four bobbin per lacer and producing decorative braid, and also a spiral braider with twenty lacers.

Steeger ② with its new design of lacer carrier bearing and guides similar to bobbin lace techniques and available in modular form, demonstrated a flexible approach to resol-

---

① Nylon inserts in the grooves of the flyer discs and roating bushes between the two lacer discs… 耐纶嵌入锭翼圆盘的沟槽及两个花边筒子之间转动的套筒……
② Steeger　Steeger 公司

ving problems. especially in the speciality machine sector. With the special design of the bearing segments, braiding machines can be built with any desired numher of flyer discs, which is especially advantageous for applications in the technical sector where lacer numbers can exceed one hundred.

Trenz Export[1] is now also offering braiding machines with safety cladding of machines.

Wardwell[2], a maker of high-speed braiding machines especially for wire braiding, offered a fully encapsulated machine of conventional design for lower noise emission and increased safety.

Bobbin lace machines

Hacoba[3] developed a bobbin lace machine with electronic yarn withdrawal. The bearing and movement of the lacers are based on tried-and-tested principles.

The O. M. S. bobbin lace machine has improved yarn withdrawal with a coated withdrawal roller surface[4]. A new electronically controlled central lubrication system appreciably improves maintenance of the machine. A further innovation is a heater element downstream of withdrawal from the lacing point, which melts fusible bonding threads at 80℃. Significant noise reduction has been achieved by encapsulating the lower part of the machine. The new lacers of the machine have been designed in such a way that there are now no guide rods subject to snagging on the outside of the lacer[5].

Winders

It has been noticeable in recent years that increasing number of braiding machine makers are getting involved with the development of automatic winders, since it is widely recognized that braid quality is very much influenced by the quality of the lacer bobbin.

Hacoba developed a four-spindle electronically controlled automatic winder of new

---

① Trenz Export　Trenz Export 公司

② Wardwell　Wardwell 公司,产品有高速编带机

③ Hacoba　（德国）哈科巴公司

④ The O. M. S. bobbin lace machine has improved yarn withdrawal with a coated withdrawal roller suface　O. M. S. 梭结花边机通过退绕罗拉有涂层的表面改善了纱线退绕条件

⑤ ... threr are now no guide bar subject to snagging on the outside of the lacer　……现在在花边筒子的外面没有易受损的导杆

ergonomic design. Luigi①, the Italian subsidiary of Hacoba, developed a new automatic winder capable of producing precisely prescribed bobbin lengths with push-button control. The thread end is automatically started on the bobbin by means of clips. Herzog offered an electronically controlled four-spindle automatic winder with automatic variation of bobbin length.

Winding machinery makers too developed speciality winders for lace bobbins. Cezoma②offered winders of new design with electronic control accepting up to twenty programmes and undertaking a variety of monitoring functions. To maintain constant yarn lengths thread tension is measured by a sensor arm which controls the downstream thread tensioners.

O. M. R.③developed a four-spindle fully automatic machine with electronic drive for producing parallel-wind and cross-wind. Package build is controlled by a liner motor. Offered for the first time, the new "RA-4" four spindle fully automatic winder by Ratera④ produces a bobbin up to 200 mm in length.

Vocabulary

法厅 knock-off 终止,停止(工作)
harden (使)坚固,(使)结实
wire 金属丝
fusible 易熔化的,可熔化的
snagging 擦毛,磨损
clip 夹钳,压紧,布铗
parallel-wind 平行卷绕

bush 套筒
cable braiding machine 编绳索机
lacing point 编织点
guide rod 导杆
subsidiary 附属机构
lace bobbin 花边筒子
tensioner 张紧装置

---

① Luigi   Hacoba 公司的意大利子公司
② Cezoma   (荷兰)塞泽玛公司
③ O. M. R   O. M. R. 公司
④ Ratera   Ratera 公司

Think or answer these questions

1. What use does nylon insert in the grooves of the flyer discs for?
2. What does the significant noise reduction of the machine by O. M. S. have been achieved by?
3. Is a new automatic winder by Luigi capable of producing precisely prescribed bobbin lengths?
4. What is to maintain constant yarn lengths thread tension measured by?
5. Which wind shapes is a four-spindle fully automatic machine by O. M. S. for producing?

Part 2 Translations 参考译文

第1课 短纤维纺织(一)

在短纤维纺纱方面，近年国际纺织机械展览会基本上没有推出新工艺或设备。这种工艺的改进，与提高生产率和质量的问题相关。在工业中，自动化活动依然停滞在已经获得的高标准程度上。工艺数据的收集和生产控制，仅在经济发达地区成功地保留。另一方面，工艺与技术设想已享受到一种复苏，使它们能对改进质量、更高的生产率、降低维护需求及改善能源利用方面有所贡献。

来自于1991年国际纺织机械展览会的一些新工艺也已经牢固地确立。

清棉间

现在，实际上自动搁置棉包之后，要斜角度地自动抓包是可能的。在自动拆除捆包物和手工除去包布以后，借助于自动加载小车，将斜着工作的抓棉机构加载。高度的灵活性为若干抓包机系统服务创造了条件。整个系统是积木化设计的，能够一步一步自动执行。

抓包系统的智能水平，由于电子计算机的控制得到了改进。用于自动批量装载的设备，从垂直放下到倾斜放下，且结束时而没有任何残留纤维。在开始混棉之前，不再需要调整不同高度的棉包的水平位置，其次，抓棉机构系统的需求(生产速度)靠抓棉机构的增量自动控制。抓棉机构系统的速度调节用频控电机装备，使快速适应卷入的纤维无需变换传动装置成为可能。

今天，像这样的抓包系统的生产量已达到1600 kg/h。一些混纺能由一个带转塔的抓棉机构进行。控制与故障探测装置在抓包机上都备有。

在清棉工序中，串列布量(3~4个罗拉连续地)装配的罗拉清棉机的概念已经被普遍接受。就清棉效率和限制损伤纤维而论，在相对低的罗拉圆周速度下，质点的高离心力的清棉原理似乎是最宜条件。适应于纤维的需求是通过锡林上的针布来实现的。

像特铝茨勒尔公司那样，马佐里公司具有修改清纱水平的方法，通过伺服电机无级调节清纱器刀片来完成。除刀片清纱器外梳棉板也可用于开棉目的。力达公司继续支持不直接握持纤维的非损伤清棉过程及用计算机控制清棉工艺强度。所有的系统都提供快速适应对特殊原料及清棉要求水准的机器调节。在清花间后面，关于异纤维的检验已经做了艰辛的努力。与强大的计算机技术结合的现代图像处理设备，使得以差错、发射或反射光产生的连续图像的快速分析成为可能。这不仅对不同颜色的颗粒，而且对不同反射特性的颗粒的检测也可行。确保最小纤维损伤的更进一步的特点，是具有50 ms反应时间的快动作密封板装置。

在清花间检测异纤维的观念，似乎已形成纺纱厂的一个基本特色。现在，具有最少打击点的短流程清棉生产线已经变成现实。

第2课　短纤维纺纱（二）

粗纱机

目前，在质量与提高生产率方面最重大的技术进步，将在梳理领域中发现。例如，在运转部位梳理机盖板的数量在减少，并且被固定盖板所取代。带弹性针布的回转盖板，现在的主要功能主要是除掉短纤维和残留杂质，解办纤维，去除棉结。

开松是由上游的固定梳理部分或盖板来完成的。这些带有金属针布的梳理组件，能比弹性盖板针布完成更高的工作负荷。现在，梳理机制造商已经为他们的产品研发出定做的组件。所有的梳理组件，目前明显地比它们过去具有更高的精密度和更大的调节可能性。这些(方面)使更多有差别的可重复的调整，在长期的运转中成为可能。在梳理部位，这些方法已经能够使回转盖板的数量减少到21根。空心剖面的盖板比铸造盖板在悬垂性能上有更大的精确性，而且梳理性能在机器整幅范围是恒定的。

特吕茨施勒尔公司的三刺辊系统排列让一种旧想法获得再生。由于愈来愈大的生产率的压力，要求采取措施在锡林盖板区梳理部件上减小负荷。随着纤维毛茸非常充分地现出网状，纤维离开刺辊区。采用这样的装置，梳理机的生产速度超过140 kg/h具有良好的质量现实的。而当一种原料给定而且生产速度已知时，梳理质量的提高导致更好的能源利用，梳理机针布的有效寿命也会延长。

三刺辊的排列，是朝向在梳理机上长期更高生产率的方向走出的合理的一步。梳棉机因此格外含有了清花间的功能，因为当给出吞吐量时，这种三刺辊排列能够突出地经过可调整刀片，单独执行除杂功能。

但这种高效率的梳理机，现在比以往更需要可控制的重复调节。对锡林盖板区有适用于盖板精确调整的装量，在不同的设定场合用盖板控制或用由邓肯通夫研究所研制的疏理工艺的设定系统。这种调整由电子感应器显示。报道的装置保证，在重复使用或磨合以后，借助其初始设定，梳理机可以重新设定。因此设定是不受主观因素的影响的。ITV 系统可以适合于任何梳理机，锡林的同心度与匀整也能被测定。特铝茨特公司也已经开发出对全部产品标准的在线棉结计数。现在，梳棉质量的客观监测是可能的。

（英国）克诺思诺的双联梳理机继续代表着三刺辊的发展。每根盖板在高度上可以单独调节。这就减少或排除对成套盖板研磨的需要。硬的尖端被维持，延长了使用寿命。

因此，现代梳理机的设计，借助提高纱线质量和增加生产率，使能源得到更好的利用。更好的能源利用，为在梳理方面的资本投资提供了最经济的生存基础。

（美国）豪林沃思公司展示了一个仅仅使用静止梳理部分的梳理系统。这种新的疏理机的清洁系统包含 6 个窄的带刀片的梳理组件的系统，在针布和设定方面非常灵活。实际试验肯定，该系统现在也能够成功地用于棉处理。未来将证明，不管是局部还是全部排除带弹性针布的回转盖板实际上是可行的。

并条机

简单的三上三下牵伸机构或许是最广泛的组合系统，从而成为适合短纤维处理最成功的牵伸构造。新型 Truzschler 并条机采用多级电机驱动，用于包括匀条的进出。机械传感被保留但用沟槽集棉器。现今的自调匀整系统和快速反应伺服电机，如条子断裂发生，允许校正长度大约 10 cm。矩形条筒显露出可以取代圆形条筒。从转杯纺纱出发，现在开始考虑粗纱机的换条筒和为喷气纺纱喂料。较多的空间利用和在自动化场合更易操作，在最后的并条阶段要得益于矩形条筒。并条工艺过程靠连接系统能完全自动地联合起来，而且并条速度已实际实现 800 ~ 1100 m/min。

精梳机

CSM 推出一种体现技术上重要新思维的全新精梳机。用于所有八精梳头的集中重复喂入调整，不必变换齿轮组合使操作更容易。用气流加压的灵活的下钳板，自动地适应棉卷及棉卷中棉条形成的不匀率。与这有关的是在分离时改进的纤维控制，其允许喂入棉卷的重量达到 100 ktex。协同操作的分离罗拉与精梳边缘气流维护改善了条子接头。靠自动操纵运输机，从条卷机自动喂给精梳机是可能的。速度达到 500 钳次/min 似乎是完全可行的。速度达到 400 钳次/min 一般是可行的。力达公司仍然拥有唯一带全自动棉卷供给的精梳机。

YARN AND FABRIC FORMING
现代纺织英语

第3课　短纤维纺纱(三)

粗纱机

　　显示出最先进的粗纱机设计,是高诺森海纳公司的 BF224 型棉纺粗纱机。除牵伸变化外,多电机传动不断取代所有的机械传动。锭翼距变到 224 mm,在机器上排除了常规的间距间隙。一些粗纱机具有条子感应控制机构。

　　高泽公司粗纱机是另一种运转方式,没有变换齿轮组,而具有机械/气流控制组件。像圆锥轮卷装成形这样经过实验验证的机械组件,具有高精密调节装置。所有的粗纱机不仅具有络卷自动选择功能,而且具有喂入卷装再补充的自动选择功能。

环锭纺纱

　　在所有的纺纱阶段,环锭纺纱近些年浓缩的发展是最显著的。具有达到 1632 锭的超长机器,在锭子驱动方面涉及到新概念。与成功的四锭分段传动系统一起,用主轴驱动和窄皮带传动的分段传动也出现了。就绪森公司系统来说,分段的皮带驱动 48 个锭子而且是环状带,取代目前的系统没有问题(更新换代没有问题)。高泽提出一种新的传动概念。即使在长机器上,机架也仅有一条龙带,由七部电机分部传动。这条龙带于是只有一种速度,而且能使这条皮带很窄,导致非常低的噪声水准。牵伸机构的轴承能够沿长度方向精确无级别地调整,为在长机上安装罗拉时提供极高的精密度。

　　由 ITV Denkerdorf 开发的 Cerasiv 陶瓷纺纱钢领,在环锭纺纱中代表着一种创新。这种纺纱钢领,排除了以现今钢丝圈速度进行的、昂贵的钢领试车程序。其次,特意研发的钢丝圈,延长了使用寿命,而且减少了钢丝圈的频繁更换。工业上的试验证明,钢丝圈的寿命比在目前普通系统里的寿命长 2.5~3 倍。该系统在精纺毛纺纱中,不需要润滑管理。在纱线质量、飞花、维护及寿命方面,这里有重大的效益。这种陶瓷钢领的使用,在国际纺织机械展览会上,是在 CSM 毛纺环锭纺纱机上演示的。在环锭纺纱的所有领域,现在关键是锅丝圈磨损成为通向更高生产率的一个障碍,这个系统标志着在增加生产率方面的一个起步。

　　没有传统纺纱三角区的纺纱(紧密纺纱),通过更高的纱线强力,使更好的纤维利用成为可能,捻度水平也能被减少。这种纱的一个特点是它的低毛羽,而且钢领/钢丝圈系统不需要润滑剂。在关于紧密纺纱装置的初期经验方面。像在

CSM 已经演示的那样，就毛纺纺纱来说，以陶瓷为基础的钢领和钢丝圈系统，正在扮演一个重要的角色。

非传统纺纱系统

在转杯纺纱方面，近年推出了基于 Elitex 研制的众所周知的纺纱箱。具有导向轴承的纺纱杯的速度增大到 90000/min。高速转杯纺纱机仍然是赐来福公司和力达公司产的带非导向轴承的机型。现在赐来福公司获得了接近 150000/min 的速度。与 1995 年国际纺织机械展览会比较，相当于增加了大约 20%。转杯纺纱机的性能限制，在中支范围更取决于卷取速度。立达公司产的 R 型转杯纺纱机带有纤维流反向的纱线接头系统，使连接成为可能，明显地提高了织造生产效率。一种具有极大灵活性的全新纺纱箱-SE 10BOX 由绪森公司推出。全部纺纱组件能够容易地互换，无需任何特殊工具。使喂入率适应转杯直径操作时，无需取下面板。能源消耗已经减少，随着打底筒子的继续应用，SE 9BOX 的普通自动装置已被采用。

以 400 m/min 的速度，在所有的短纤维纺纱系统中，喷气纺达到最高的纺纱速度。一种新推广的机器，具有单喷再加上用于产生假捻的正交摩擦罗拉。这些罗拉产生出非常低毛羽的织造用纱线，具有至少 40% 聚酯含量的混纺纱也用 804RJS 来纺。可以用接头装置供给这些机器，作为打结器的替换物。在这种场合，从纱线卷装出来的纱头被放在抬起的罗拉上并以全速接上头。喷气纺也是为了稳定地求得更高的生产率。喷气纺是否将达到灵活的理想水准，尚待观察。

第 4 课　化学纤维的初纺（一）

长丝纱

近年来，出现非常多的合纤长丝机械供应商。随着对机器设备的不断改进，有一个时期这些机器只适合于聚丙烯，许多技术公司推出了能纺大多数聚合物的设备。

现在，有两种主要的机器似乎是节能与工艺间断最少的。这导致了大部分公司选择下装式喷丝头组件，它避免了上装式引起的与烟囱效应有关的热损失。工艺间断被减少，例如，通过连续过滤器的更好的聚合物过滤，及装配无间断或任何损失落筒的转盘式卷绕机。

尽管用于生产未拉伸丝（通称为低取向丝或 LOY）的卷绕机可获取，但更多的

YARN AND FABRIC FORMING

重点被放在 POY、HOY、(高取向丝)和 FDY(全拉伸丝)生产线上。

——交缠喷嘴

许多公司都强调,适合于他们生产线的交缠喷嘴的高效率。这减少了生产的间断,给予了更好的卷装成形,更重要的是减少了用于连续运转要求的尺寸量。这有益于环境。

——卷绕机

卷绕机的发展似乎已经直接对准了更长的筒管夹头,它能卷绕更多或更宽的卷装。典型的是 Barmag 的 CW1500 型卷绕机,它具有 1500 mm 夹头长度,能放 6/S、8/S 或 10 个筒管。这种卷绕机生产 POY,速度在 1500~4000 m/min 范围。

——POY

作为 Automatik GmbH 在 1992 年发现的结果,Rierter 能在国际纺织机械展览会上第一次推出熔体纺丝机械。

特别有趣的是,一种具有象把纺丝泵安在纺丝箱体上面,把联苯锅炉安在纺丝箱体旁边改革特点的 POY 机。后者,与其他的节能步骤一道,据说减少 25% 的热损失。

——FDY 纱线

约翰·布朗德国工程技术公司作出了一项为生产全拉伸丝的非常有价值的开发,传统的 FDY 机用导丝辊来牵伸纱线,而 John Brown 热管拉伸机靠在热空气逆流中加热长丝,把取向引入聚酯长丝。据称该工艺可以获取更低的成本、更好的均匀性、较少的断裂和均匀上染率。卷绕机的使用速度在 5000~6500 m/min 范围。

另一种非传统 FDY 制法是 EMS-Invensta 的 H4S 法。在这种工艺中非加热的导丝辊拉伸纱线,而且一个蒸汽室同时交缠和松弛该长丝。

——工业用纱线

由聚酯和聚酰胺聚合物制的工业用纱线,是在一步法机器上生产的。对于普通的聚酯纱线,Barmag 推荐的速度在 400 m/min 范围内,而对高模数低收缩(HMLS)聚酯纱线(现在用于轮胎上),使用的速度在 8000 m/min 范围内。

另一方面,对于最高强力的聚丙烯工业纱线,UK-based Extrusio Systems Ltd. 和 PFE Ltd.,都推荐采用二步法工艺。据称 90 g/tex 范围的强度,在许多应用上使聚丙烯成为聚酯和聚酰氨的一个竞争者。

第5课 化学纤维的初纺(二)

——BCF

三股纱地毯机,在许多地方都能见到。Barmag 声称,它的三股纱机取得了空前的/性能比率。Plantex 推出一种新的高速 BCF 小型设备,以 3000 m/min 范围内的速度,生产从 1000 到 3600 丹尼尔的三色纱线。该机能处理所有类型的可熔纺聚合物。

——纳米纤维

据报道,东丽公司已经研制出世界首例纳米尼龙纤维。这项技术适用于高聚物纤维(如尼龙、聚酯),并能在常规设备上生产。

短纤维纱

短纤维生产线分成两种类型,二步型和紧凑一步型。这种更新的紧凑型机器,在一个连续运转中,从提供小颗粒到打包,包含了所有的工艺步骤。

Fleisaner 认为,紧凑设备非常适用于纺前染色的纤维,因为颜色的色泽能被快速检验。Heissner 和 Neumag 都声言,在这样的机器上甚至能够经济地生产小批量的短纤维。

Fleissner 公司除了提供用于聚酯-聚酰胺及聚丙烯生产线外,还提供湿、干纺聚丙烯生产线。

今天,适合于短纤纺的每个喷丝头组件上细孔数量,典型的有数千个,像依姆斯-因维塔公司的短纤维生产线上的有 6000 个。由于喷丝头弄得较大,所以纺丝头组件变得更大更重。在 Zimmer 看台上,见到一个半自动处理装置用于自动集中自动封闭的组合件,据说重 250 kg。这家公司也研制出一个新的"大卷装"卷曲箱,只用一个卷曲箱就能使一个单独的,日产 200 kg 的生产线运转。

过去经常叫做"新型"或"差别化"纤维,不再是新型或差别化产品。一些公司现在推出纺丝组件/纺丝头来制造这样的纤维。例如,Neumag 推出设备用来生产并列和纱芯/外鞘双组分、空心纤维,甚至空心双组分纤维。

特别令人印象深刻的双组分技术,由 b. g. plast 开发。一台紧凑型机器正在纺聚酯双组分短纤维,据说可能是通过一个含有 20000 孔和 30000 孔的喷丝头组合件。

东丽公司提供生产海岛型复合超细短纤维的生产技术。

结论

如此之多的技术设备公司,正在推出用于生产化学纤维的优良机械。事实上，在这样一个广阔的范围，让潜在的客户作出决定非常困难。因此，许多公司的小册子都强调，公司在该领域的长期经验。这对想要购买的厂家来说，是必须要慎重考虑的。

另一方面，人们只能意识到交货时间、技术帮助、担保，而价格将是最主要的因素。

第6课 纱线的加工、包装与处理(一)

在论述加捻方面的新产品和讨论在国际纺织机械展览会上注意到的几个动向之前，我们将简要地指出在纺织工程所有分支中的明显趋势。换句话说，细节的改进措施，广泛使用微处理机，用于络筒机头和锭子的分别传动，在光驱(CD-ROM)上增加数据记录及监控环，操作手册的目的之一是加速机器故障的识别和故障排除，改进自动装卸，以及不惜一切代价实现工艺自动化。

并线络筒机

并线络筒机的走向，朝着连续精密卷绕的方向发展。这样导致更高的排列密度和更高的卷装重量。在喂入倍捻机的纱线中，仍然有单根丝线与再捻络筒卷装之间的竞争。赞成或反对这两个系统的争论依旧存在。实际应用必须在仔细分析两个系统的优缺点之后作出决定。

包芯纱和花色纱

在包芯纱与花色纱方面，除了细节的改进没有根本的创新。为控制机器及其运转而渐渐引入微处理机，使实际上无限制的设计机会成为可能。

适合于销售点包装的卷绕机械

适用于销售点的络筒机的主要特性，是它们自动化的高水准，近年来其仍然在进一步地发展。精确的长度测量，具有电控系统的简易操作和用于辅助用途的装置被认为是理所当然的。销售点的包装，包括许多不同规格和类型的锥形和圆柱交叉卷绕卷装，加上绕球机和用于生产绞和绞纱的机械。由SSM推出的一种新型机器是Preciflex交叉卷绕机，适用于卷绕长丝纱。

用控制单元，实际上通过一个触摸显示屏给所有的纱线卷装参数集中编程是可能的。这种自动化的高标准能使所有想得到的卷装类型和形式容易生产出来，不仅以销售点形式，而且以其他形式例如适合于染色的卷装。

连续处理

在加工中的纱线的连续处理(像定形和膨化变形)方面,没有任何重大变化,尽管一定的变化在各自的机器中能见到。

加捻系统

今天的加捻系统包括环锭捻线、倍捻和无气圈加捻。无气圈加捻过去常被称为二步法加捻。环锭加捻的重要性仅停留在少数特别的场合,像前面提到的花色加捻。倍捻捻丝最近几年渐渐地流行起来,现在它有一个值得注意的竞争者,在接下来提到的新机器中。

在1995年国际纺织机械展览会上见到的一个真正的发明,首次由Hamel(Saurer集团)展出。这就是第一次于1993年推出的"Tritec-Twister"三倍捻线机。在原理上它是一个无气圈捻线机源于普通的Hamel 2000 二步加捻机和具有二个嵌套逆向旋转纱罐的上行捻线机,实际上获得了锭子转动的三倍速度。换句话说,锭子转一转,在最终纱线上加三个捻回。

像在原先的机器中那样,由于在纱罐中线摩擦非常低,所以丝线受到的应力小。与更高的生产率一样,制造者声称其优点是低能源消耗。机器中装有自动空气穿线机构,由一个脚踏板驱动。

第7课 纱线的加工、包装与处理(二)

倍捻机

在倍捻机方面,根据市场的需要存在一种对两种类型机器需求的趋势。用于高容量生产,有一种更简便的窄形机器,专门地用于纱线加捻。另一种类型的机器,具有更复杂的技术,供更多通用的使用及特殊产品。

在倍捻方面,在1991年国际纺织机械展览会展出的,把自动化络筒机或并线络筒机与倍捻机连接的系统,近年来不再展出。在它们被使用期间,自动化操作筒子变换器,即适用于用其复杂的穿经工艺给倍捻机换筒,又适用于落筒满卷的卷装。即使在那段时间里,也不能肯定地断言这样的连接系统的经济可行性。工业已经不接受这种形式的自动化。另一方面,供料卷装、捻线卷装和空管的自动装卸已完善,而且穿纱工具也可用。

汽蒸和给湿

纱线卷装的汽蒸和给湿,从本质上讲并不是一种新技术,而且除了少数细节改进，近年来没有变化。Welker 和 Xorella 在该领域都出了研究成果，肯定增加棉纱含湿量的好处。例如，Welker 在一个织机加工的收缩试验中，能够证明纱线的性能得到了极大的改善。Xorella 受委托的棉纱物理性能的研究，从优质的纱线强度和伸长特性得出结论，在下游工艺中有好处。因此，纱线的汽蒸与给湿比简单地给纱线增重具有更大的重要性。该系统的经济可行性，需要考虑在质量提高方面从而改进纱线性能，其次增加重量。在经济分析中，当然需要包括汽蒸的能量消耗，这取决于所使用的系统。

总之，在这个特殊的领域，可以随着目前的发展趋势及近年来技术装备的开发，工厂将被更好的装备以完成他们的工作。

第8课 变形与长丝纱加工（一）

拉伸变形机械

具有越来越高生产速度的拉伸变形机械的发展的结果，是加热器及骤冷装置变得更长。它导致了宜接近性受到高设备的限制。这种趋势近年来随着短高温加热器（HT-heater）的采用，得到了缓解。所有的机械制造商，在每台卷绕络筒机头都提供自动化落筒。可预计这些发展产生了新的设备配置。

——在初始加热区域中的高温加热器

在初始区域中随着高温加热器的采用，加热区从约 2.5 m 减小到不足 1 m。高温加热器中的温度在 300 ℃ ~ 500 ℃，定形温度大约为 200 ℃。加热器长度、加热器温度及丝线速度必须相互平衡，通过这种方式在加热器出口处精确地达到纱线的定形温度。采用更短的加热器，新的设备外形结构使机器的高度更低成为现实，同时具有更易接近性。另外，由于改进绝缘材料及减少辐射表面，达到能源节省。

——集成的自动化落筒

到目前为止，卷绕络筒机一直位于机器的中心，即紧接地在第二个加热器的前面。用这种方法，从一个过道看管机器一侧是可能的。现在，一些机器制造商正在推出具有改进外形的机器。包括自动落筒的卷绕装置被放在看管过道的相反侧。这意味着一个看管过道加上第二个用于纱筒取出及空管供给的过道，适合每个机器侧面。由于这种机器外形占用更多的空间，所以一些机械制造商保留了前者的外形，每台机具有一个侧面过道。可以假定，集成的自动化落筒及每台机侧必然也有两个过道将变成可以接受的现实。长丝的穿线方法多种多样,解决的办法既简单又复杂。自动化的发展仍然存在。

适用于生产高弹性纱线的单加热器和喷气膨化器,由于成本的原因由对应双加热机器的部件获得。

——摩擦加捻装置

在摩擦加捻装置中,9 mm 厚的圆盘已变成标准做法。圆盘材料的选择取决于变形纱线的最终用途。对加捻部件单独传动的问题已被解决。这种发展已显著地降低了噪音水平(大约 12 分贝)。这种系统的经济性还尚待观察。Temco 推出了一个装置,它能从 Z 捻切换成 S 捻。另一些制造商研制了用于三维加捻机元件的能互换的组件。这些新装置有助于变形工艺更具适应性。Muratec 采用了一个装置,它同时生产邻近的 S 与 Z 捻丝线。这两根不同捻向的纱线被互相混合并卷绕起来。这就形成了一根变形的长丝纱线,没有缠结倾向。在变形处理之后,加一个喷气相互混合来改进退绕性能是可能的,并可能排除上浆的需要。

所有的机械制造商都采用在线测量,通常张力测量头用来监测变形过程。

第9课 变形及长丝纱的加工(二)

趋势

在织造加工中可以看出以下的一般趋势:

——更低的结构及更易接近

——节省能源及更低的噪音水准

——更具灵活性

——在线质量监测

——变形工艺在全面自动化方面的应用

全取向非变形纱线,一方面能由 FOY 或 FDY 工序纺丝制成,另一方面可以通过 POY 辅助拉伸工艺(拉伸—卷绕,拉伸—加捻或拉伸—整经)来生产。

拉伸—卷绕和拉伸—加捻变得更重要

提供拉伸—加捻和拉伸—卷绕机械的公司数量,1995 年与 1991 年比较已进一步增多。这些系统适用于小批量和特殊纱线。拉伸—加捻和拉伸—卷绕机械,到目前为止主要由化学纤维生产商购买。建议今后纺织加工商也要更多地使用这些设备。这将能使纺织生产者以最好及最灵活的方式,使拉伸工艺适应自己的产品。POY纱线在全世界都能以低价格买到。

Zinser提供的"Co-We-Mat"自动落筒适用于拉伸—加捻机，能够在拉伸—加捻和拉伸—卷绕机上生产变性长丝纱线。差异收缩纱线是不同收缩率纱线的混合。异染色纱线由两种或多种不同染色亲合性的纱线组成。粗细纱更具可能性。纺织生产商要开发差别化纱线，这是一个机遇。在经用纱中，可以使用POY纱线，并由整经拉伸工艺将其拉伸。少数整经—拉伸—上浆联合机已经组合成，但没有获得广泛的成功。相反，人们正在偿试使用一种缠结处理，对长丝纱进行上浆的工艺。整经拉伸是否会获得广泛承认，尚待观察。

加捻机械

在加捻设备方面，除了花式加捻机、倍捻机之外，搓绳机正在不断地受到欢迎。搓绳机不仅适用于轮胎帘线和地毯纱线，而且在更轻的结构方面适用于缝纫线和粘胶纤维的刺绣纱线。芯纱加捻、纱线包覆和喷气包覆对弹性纱线的重要性越来越大。

Saurer的"Tritec-Twister"三倍捻机正在不断地得到普及。在加捻机械方面，很显然专一性机器正在取代通用性机器。加捻设备的完全自动化并不是真正地经济可行的。近年来，推出了局部自动化的设备，这是有成功前景的途径。

第10课 毛纺机械（一）

近年来，就精纺、半精纺及粗纺毛纱工业而言，没有人想到会有技术与工艺上的轰动。上升的成本加上国际竞争，正在迫使机械制造商在一个适当的时间周期内从经济可行性和市场接受方面，考虑他们的新发展。这些因素导致了若干对策，其中（包括）使经过试验及验证的机械设计正在适应性地更新，合理的技术让步被接受，用于系列生产，并通过合理化降低成本。然而对于精纺、半精纺和粗纺毛领域来说，一些重要的特点是显而易见的，即：

——试验技术的工艺的现状及稳定性
——通过对机器组件的空气供应的改进，来改进与机器清洁度相关的细节
——在驱动及润滑技术方面的改进
——机械零件尺寸的优化
——单独机器元件的创新，尽管其实际意义尚需长期试用来证明
——数据记录、分析及存储方面的改进
——在操作及人类工程学方面的改进

用于精纺和半精纺纱的机器——梳毛

有效的机器宽度被限制在 2500 mm 以内。由于大锡林直径增加到 1500 mm 加上更高的圆周速度，生产率与老式的有效宽度超过 3000 mm 的精纺毛机相等。一项新的改进是用双道夫出条和改善吸尘装置。新道夫斩刀每分钟提供达到 3600 次的动作。梳毛辊（罗拉）相对于锡林间隙的远程调整使精密设定成为可能，包括后面的碾磨。

——精梳机

所有的精梳机制造商都已经把速度从每分钟 200 钳次增加到 250~260 钳次。进一步值得注意的改革包括：

适用于一级精梳毛的双头精梳机。

毛网限制和净毛条形成归因于在皮圈托持板上可调低压空气压缩机的作用。

一项引进是回转梳刷喂入梳栉"Rotat Feed Comb"。用这个装置，连续保持清洁是可期的，同时噪音比传统喂入装置更低。

——牵伸机构

在精梳毛牵伸机构方面没有显著的改进。校平器技术已经从机械发展到电子技术。伴生的益处——更容易编程、校正范围更宽（±25%）及数据记录，对于减少换批时间，特别在小批量的场合是重要的。一项非常有趣的改进（其减少了机械传动组件的数量）是用 4 个变频电机控制机械传感和电子毛条卷控制。

——高牵伸搓条粗纱机

这种普通的高牵伸搓条机继续被使用而无明显的变化。

——精梳毛纺粗纱机

BF 三道粗纱机（有非常陡峭角度的牵伸系统）已经填补了一项技术空白。靠牵伸机构输送的粗纱能在同一角度和同一长度方向被送到前、后翼锭。粗纱张力及捻度水平是恒定的，即使对纤细纤维也一样。

——精梳毛细纱机

"Plyfli 2000"采用了精密的细纱机形式。与传统的经过改进的装置一道，包括自动断头修补、带纱线测长的自动落筒、电子清纱器及连接加捻工艺，适用于 Plyfli 的包芯纱附属装置是特别值得一提的。输出速度高达 280 m/min 可获取。这种技术现在完全已被掌控。

第11课 毛纺机械(二)

——环锭细纱机

环锭细纱机 RKW 的特殊特征是上下皮圈的异常长度。下皮圈由两个下罗拉传动，而上皮圈由两个中间罗拉接触摩擦来传动。该系统取名为"Toyota'sOrignalb CRADLE 的四罗拉系统"。已经申请专利保护。

紧密纺纱精纺毛工艺声具有毛羽减少、纱线强度增强、纤维利用率高、加捻底、锭速高、纤维损失少、生产效率高等优点。在控制工艺方面的问题，需要通过持续的试验来解决。

Fiomax 2000 环锭细纱机，采用 17 mm 直径的锭盘结合的 HP-S68 锭子轴承，具有超过其他锭子传动系统的优势。节省能源高达 20%，锭速变率低于 0.5%。频控驱动能任意缠绕，因而在生产中，完全可以控制各种质量的变化。包括最显著的是能控制不到两分钟的落纱。简单设计的和操作模式的成功改进，就是 CutCot锭盘清洁系统。要停止随机器运转的锭子，可选用固定式手柄杆闸或插入式制动器。

在纺 20~60 Nm 纱线范围内，使用回转钢领，据称生产率可提高 30% 以上，断头减少，钢领圈寿命延长到 2000 小时以上，而且钢领润滑减少。

——环锭细纱机牵伸系统

介绍一下 PK600 气体加压摇架。气流加压或局部压力卸荷的调节集中进行。新 NASA UH-56 锭子使得下皮圈变得简单，从而大大降低了维护成本。

——锭子

众所周知，HP-S68NSAS 锭子的好处值得一提，随着锭速的增加，轴承的在锭管上的载荷几乎平行地加载，一个尚未充分认识到的发展是在所有的锭速范围降低噪音 5.5~6.0dB(A)，并且与有刚性轴承的锭子比较振动明显更低。

——钢领/钢丝圈系统

Ceratwine 是钢领/钢丝圈名字的统称，其具有 100% 的陶瓷钢领和一个经过适当设计的覆盖陶瓷的钢丝圈。它在精纺纺毛中的采用，需要长期试验和经济可行性分析的支持。

半精纺纺毛机械

OE005 开放式转子纺纱机，是为纺纱支数在 0.9~5.0 Nm 范围而设计的。当与有过一次牵伸处理的粗纱条子协力使用时，其益处是最受称道的。在转杯直径达到 150 mm 时，转杯转速可达到 1500 rpm/分。

用 Dref 2 和 Dref 3 型摩擦纺纱机，可替代半精纺和粗纺环锭纺纱机，纱线数可达 10 Nm。这些机器在某些细节方面已经提高了等级。

半精纺环锭纺纱机几乎没有变化。现在毛条自停装置、半自动或全自动落纱装置加上数据记录均是标准化的装备。高性能机器和相对低的需求，意味着重要的新模式不大可能出现。另一个因素是，在市场的主要部分，它已经被经济上有吸引力的包缠纺纱工艺所取代。

普通的 Robospin B6 走锭纺纱机，含有自动落纱和搓条筒管置换，据说仅适用于细支粗纺毛纱及高质量的原料。

近年来，改进与创新在广泛的范围是显而易见的，但往往只是细节而已，它们包括：

——传动系统的改进与修正

——润滑系统的改进与简化

——在适合维护的牵伸机构的易接近性和换批操作方面的改进

——具有适于机械状态和设定、产品质量及经济可行性的可重现参数，适用于记录、保存和分析的有意义的可编程系统

——用于机器清洁的气动系统的改进

——由于成本的压力，放弃不经济的改进

机械制造业的供应商开发了一些非常有价值的创新，但它们的经济前景及实际成功有待于持续的工业试验来证明。它们包括：

——紧密纺纱

——气压摇架

——新 UH56 下皮圈托架

——Rotat Feed-Comb 回转梳刷装置喂入梳栉，适用于精梳机

——Ceratwine 陶瓷钢领/钢丝圈系统

——回转钢领

与预期相反，要机械制造商从工艺技术方面满足所有纱线生产者的要求是可能的。但是，事实上对他们的专业产品范围而言，在该领域的公司需要的实际上是通用机械或专门化机械，而这两者都不是经济可行的。在一定程度上实现了较短的停顿时间和较快的批量更换。但是，由于技术和物流的限制，工艺的连接在经济可行性方面仍然受到阻碍。

第12课 空气工程(一)

空气调节

从经济与性能的角度出发,除了现存系统的最优化外,还有一个明显的趋势就是,在纺织领域中空气调节系统的主要制造者,从空间空气调节转向单机专业空气调节。这源于现代高性能机械设备对空气工程的要求发生了变化。由于热源和内部气流的存在,温度和湿度分布的变化可能出现在有关工艺至关键部位,从而影响生产效率与质量。传统的空调设备不再能以合理的成本消除这种不正常现像。

空气过滤器

在排气装置空气过滤方面,与吸风除尘和废料处理相结合的圆盘加滤筒过滤器,已经成为公认的系统。大多数制造商已作出设计改进,以扩大过滤面积,减小总体尺寸,从而显著地提高了性能。这里被提到的系统包括: Benninger BEN-VAC、Luwa Bahnson Multi Drum Vac、Wiesner、LTG Kompakt-Filter-System、Mazzini 等。

空气洗涤器

Luwa 和 LTG 都推出了他们的新型高压洗涤器。这两个系统都含有压力在 10~100 bar 的水喷雾。液滴形式大体被抑制,并产生了一种气溶胶,其中95%以上被气流吸收。与传统洗涤器相比,减少了99%以上的化水量,并且洗涤器泵的安装电能减少了80%左右。这些洗涤器的另一个优点是,他们不用再循环水运转而仅仅使用纯净水。

从卫生的角度,这是非常受欢迎的特点,不需要添加杀菌剂。但这些洗涤器在硬度和纯度方面的确需要更高质量的水质。概括起来,可以说,这种新型洗涤器需要更高的成本,可由更低的运转及维护成本来补偿。

空气控制

在空气控制方面,像以前描述的那样,有一个明显的趋势朝向专用的和通用的空气调节系统。Luwa(全部空气控制)和 LTG(织造定向的)推出了他们的系统,适用于织造机器。在这两个系统中,空气排出口直接设置在梭口的上方。在开口区域,经纱吸湿非常快,据 LTG 的经验,经纱断头率减少了8%。这两个系统的不同在于,对于 Luwa 系统,调湿的空气流向着梭口方向,而 LTG 以60%~80%的空气交换率在开口上施加一个气流的作用,这也消除了经纱上的任何绒毛。这两个系统都间接地需要空气,通过经纱下面排气口排出。在 LTG 空气出口的可调节刀片，使气流能够以横跨经面宽度上的任何不匀温度分布都可以得到补偿配。

因为当机器停台时，空气出口自动关闭，所以没有织机或钢筘的腐蚀。Luwa 也在一台环锭细纱机上展示了这个系统，而空气出口被设置在粗纱架粗纱卷装上方。

在圆形针织方面，Luwa 开发了一个"空气罩"系统。在圆形针织机上的特殊空气出口，使一个直径超过 2 米的"空气罩"在圆形针织机的周围形成。这就防止了任何外部的飞花穿入空气帽。在空气罩下面，也产生了一个均匀的大气层。初期的工艺试验表明，在圆形针织机场合，用这种形式能将疵点的发生率减少一半。

LTG 对它的"针织清洁系统"采用了不同的路线。在这种场合，纱线喂入卷装被放置在一个空气调节的容器里。纱线卷装通过旋转的喷气区域不断地从绒毛中解析出来。飞花被吸力抽出，随着排出的空气返回到过滤器。在针织头区域，"针织清洁系统"使用一种旋转的喷气发动机，当织针移开时将产生的任何飞花吹进一个排管。这个飞花在被油的残渣污染时，是分开的。

第 13 课 空气工程（二）
巡回式清洁器

清洁领域显然是由传统的巡回式清洁器，如 Luwa、LTG、Sohler、Neuenhauser、Wiesner、Jacobi 或 Canalair 提供的。这些系统已经被更新，一些与新式吹风机喷嘴配合及装备了附加的电子系统。例如巡回式清洁器，行驶中当它遇到一个障碍时，用铰链转动或卷起吸风或低压空压机导管向上。这是连续清洁中的优点，尤其在清洁的轨道延伸到整个生产线设备的上方的情况下。与 Sohler 的情况一样，为抽吸清洁环锭细纱机的牵伸系统，还安装了辅助抽吸系统。

在高性能纺纱和织造机械中，用巡回清洁器的趋势是，有目的的直接抽取飞花。仅用低压空压机强迫排除原理上的操作系统，不再能满足需要。

在环锭细纱机上飞花的收集与排除，LTG 用其"清洁纺纱系统"探索一条全新的路线。该系统已经被开发，特别适合于现代高性能细纱机。它的目的旨在直接在产生飞花处的收集。该系统利用在环轨区内的强列切向气流。通过在锭子区域环轨上安装一个环的网（连接在落纱机轨道上），切向的气流直接通过网屏。气流携带的任何纤维都被网屏收集，而由巡回清洁器抽吸排出。

从飞花网屏吸入的纤维，在巡回清洁器中保持隔离并被分别处理，由于这种飞花不含任何异纤维，可以改线返回到生产工序。LTG 声言在实际试验中，每小时 22 kg 的飞花已从 20000 锭中被清除，而同时没有网屏，只能从地面清除 12 kg

含油污的飞花。因为该网屏可以被用铰链转向一边，LTG 报告对机械维护几乎没有不利影响。

总之，可以说现代高性能纺织机械，对空气工程及空调技术提出了传统系统无法满足的要求。

一股强烈的改革浪潮可以预期，重点首先在联合的空气调节上，其次在清洁系统上，也就是说产生飞花处的抽吸排除。

第14课　数据处理（一）

很明显，现在许多机械制造商再一次把重点放在技术装备领域。尽管如此，更多地在工厂内采用联网方面，有若干令人关注的进展。目前，对于技术数据处理和管理软件的联网系统，只有初步的方法。IS9000 系列质量标准的日益普及，包括纺织工业在内，已经引发了数据处理的趋势，在各个行业中提供生产的可追朔性。在工艺数据的收集及生产计划与控制方面，一些公司推出了有价值的系统建议。

在这方面的数据处理中，有一种明显朝向图形界面与用户友好性的趋势。

纱线生产中生产数据的收集

在纱线生产方面，Rieter 在 1995 年推出了一个数据收集系统，它可以连接自己的机器。这个系统（只限于监测功能），在其实用性与逻辑设计图形界面方面是有吸引力的。

Zellweger-Uster 也推出一些新产品。与一个新型组合电容/光传感器一道，出现了一种新软件，它也能记录非周期性的厚斑点。另一种产品是环锭细纱机上速度调节器，其靠纱线断头发生控制，由 ITV Denkendorf 研制。

Loepfe 的 Mill-master 系统的显著特点，包括自由变化的图形汇总报告以及新的预防性的监测功能。SQL 接口允许外部程序员对存储数据选择访问。

Barco 公司提供的生产数据收集，贯穿包括变形在内的所有纱线生产环节。在该领域的一项引进是监测传感器的自动设定，由一条"学习曲线"执行。

工业技术提供一种有趣的生产数据收集与商业计划的组合。通过一个清晰排列的图形界面，可以检索不断更新的生产信息。不用于其他许多系统也有可能检索综合经济指标。该系统中的一个新特点，是用于管理原料库存量的一个模块。通过这种方式，可以组成混合物，管理未来的材料需求，拟定合同并进行全面的财务分析。整个系统也可用于织造方面。

织物生产中的数据收集

Banninge & Hubscher(一家 Zellweger 联合公司)开发了一些令人关注的技术创新,包括一个"记录器",它像一个个人信息接受器,配在人身上并显示信息。这个功能取代了旧式的中央信息屏。同样新的一个经纱生产模块,它除了显示用于经纱制备,也可以确定浆纱机的速度与温度变化。

第15课　数据处理(二)

新型的"纱线管理模块"对于在织造车间内有用正在进行的或"滚动的"纱线库存管理设施是值得注意的。目前生产中,用一个两级清单代码记录和保存在任何纱线剩余物。该模块提供了追溯任何纱线使用的有力手段。

除具有非常清晰实用的图形界面布局的 Millmaster 系统外,Loepfe 还为纺织行业开发了一套全新的织物检验系统。该系统的核心是一个触摸显示屏,经过这个触摸显示屏检验织物的操作者能够承担工作所需的全部记录与检索。即不需要键盘和鼠标来操作此系统,所有记录以这样的方式设计——使它们能够非常容易地在触摸屏上执行。该项技术代表了在织物检验简化方面开拓性的步骤。

Incas 通过 Antara 系统,提供了非常全面的工艺数据收集软件。新技术特征包括一个用于织物检验的红外线遥控器和一个网络卡,它允许织物信息快速传递。这种高速信息传递在快速换批方面提供了顿外的好处。

由 Barco 为它的 PCMC 系统提供一个新特点是图形界面。该系统(适合于中、小型织造车间)在用户友好方面是杰出的。例如,随着鼠标单击列表能够自动排序列表。Sycotex 系统现在已经配备了多个新设备和图形接口。在不久的将来有更多的改进的措施。该系统的另一个新特点,是容易正确地追溯纱线批次。Picanal Omni-Terminal 内,在一年左右的时间,生产数据收集的全终端仿真可以获取。在此,如果需要的话,在机器终端执行大量检索操作是可能的。而这些操作只有在工艺数据收集的主终端才有可能的。因而可以覆盖一个完整备份目录。

生产计划与控制

生产计划与控制软件的功能范围,覆盖了订单进度、开清单操作计划和原料管理等多个领域。在这个领域 Datatex 开发的 TIM 生产计划与控制程序,能按桌面控制单独定制。它具有全面的原料需求与预订功能加上内置成本核算。另一项开发是带有图形界面的程序包。可用的生产模块已由一个新的控制站补充。该控制站能使方案在单机器水平上进行。该项目计划能人工执行或使用各种自动计划技术来执行并可随时改变。然后其结果转入到 TIM 程序里。

——可行性

Wespa 公司更新了生产计划和控制系统,在纺纱、织造及染整方面具有很高的可行性。该软件包(能在 AS/400 或 UNIX 计算机上运行)可以配备从市场、采购、原料管理及生产方面的各种模块。另一突出的特点是,具有其他工作范围像会计、工资计算、生产数据记录及许多方面的大量接口。就机器的操作者而言,图形界面目前是最先进的。这种人/机接口提供的机会易于理解,同时又能使出错风险最小化,这一点已被机器制造商被认可,并被不同厂家使用。就实际情况而言,已经取得了高水平的整合。

第16课 装卸与运输的自动化(一)

更高的生产率

作为应用开发成果的结果,生产率在大多数生产过程中仍然持续逐步地提高。不过,多少有些例外,因为生产率总体上已经达到了非常高的水平,所以增长率正在下降。

在这些情况下,花在原料装卸、批量更换和相关操作的时间,对于生产的总效率而言变得日益重要。由于这些关系,对机器间相互运输系统和机器维护自动化的需求比以往任何时候更至关重要。

昂贵的自动化

这是许多年来,机器制造商一直开发与供应自动化传输及装卸系统的原因。但不可忽视的是,不仅对相关的系统,而且对可用空间和特别现场的生产设备都需要可兼容,需要涉及相关的项目工程,因此通常成本相对高。

用于机器喂入和产品收集的装卸装置常常需要执行复杂的动作,甚至承担由机器操作人员来执行工作。有效的计算任务,现在能以非常经济的成本来进行,但可靠处理设备的复杂机制通常仍然是十分昂贵的。

成本取于工序

在任何特定情况下,所需的支出很大程度上取决于自动化的工序,在评估自动化是否值得时,还有其他因素需要考虑。加工原料的价值、工厂的工资结构是一个非常重要的因素,以及涉及质量保证有关的方面。

由于这些复杂的和不断变化的相互关系，用户及机器制造商很难预测某个自动化或装卸装置是否可接受。一些自动化装置，例如用于倍捻机补充的自动装卸系统，自1991年国际纺织机械展览会起已经消失。现在几乎每个织机制造商都提供一个快速换批系统（QSC）。

条筒、棉卷及筒管的装卸

适用于纺纱厂的连续的条筒、棉卷及筒管装卸系统也可从任何主要的纺纱机械制造商那里获取。这些装置经常来自于装卸设备的特别制造商，如Gualchierani、Neuenhauser Maschierani或U.T.I.T.。这些设备的一个特点是他们可以在不同的自动化水平上使用。

粗纱管的装卸

一些空中轨道运输系统（用于粗纱管），是这样设计的，在环锭细纱机上更换筒管可以手动地随机进行。另一方面，同一制造商也提供分段换筒系统，使粗纱管立刻在环锭细纱机的喂入位置装入。对操作的全部要求只是给喂入纱条接头。

条筒的搬运

在一些纺纱厂，条筒的搬运从并条机到粗纱机，使用地面轨道运输线。另一方面，无人驾驶的运输机车用于条筒（包括转杯纺纱机用的矩形条筒）的运输像Rieter和Schlahorst这样的公司大力推广无人运输，以及包括CSM和Vouk公司的用于棉卷的运输系统。

第17课　装卸与运输的自动化（二）

对于快速批量更换的普遍认可

目前，在织造方面自动化的方面最明显的特征是快速换批系统。几乎每家织机制造商，现在都提供一个合适的系统。在1995年国际纺织机械展览会上，甚至举行了真正的"比赛"。在比赛中，换批的时间被显示在一个能从远处看见的时钟上。但直接比较换批时间是困难的，因为条件可能区别很大。有趣的是Genkinger公司为五家织机制造商的快速换批系统提供了主车。

织造中的逐渐自动化

与纺纱相比，纺纱有大量的自动化装卸系统，而织造中原料装卸的自动化才刚刚开始。在1991年国际纺织机械展览会，Sulzer Ruti和Innovatex公司展出了一个纬纱卷装装卸系统，用该系统，纬纱卷装由一个空中运输线从仓库补给，并由一个在吊车导轨上移动的机器人放入织机的纬纱架上。这个系统现在被放弃了，因为它经济上不可行。在1995年国际纺织机械展览会中，一段录像片演示了一个由Neuenhauser Maschinenbau 与 ITV Denkendorf 公司联合研制的装卸系统。一个移动清洁器装置携带一个抓取系统，它从一条运输线上取出纬纱卷装，送到特别设计的纬纱架上自动装载。与大量的适于纺纱设备的自动化装置一道，这有可能适用于装卸的逐步自动化，能使它在各种各样的生产条件下被合理地采用。

联动装卸装置

最先进的是那种把装卸系统连到原料进度和数据处理系统，把它们合并成一个单个装置，并提供计算机集成制造（CIM）系统组件的可能。但由于CIM不在这篇报告的范围之内，故将不对它作进一步的讨论。

经济可行性的解决方案

一般来说，现在已经有了大量的解决方案，适合于执行复杂的运输与装卸工作。这些（方案）是否经济可行，取决于具体情况并需要非常仔细地分析。

另一方面，机器制造商提供解决方案，如果它能被大量销售且具有经济效益，才能持续。市场将决定哪种自动化将长期幸存。

第18课　织造准备

纱线张力器/筒子架

近年，一些机械制造商推出了在丝线张力器方面的新开发，这些张力器具有低的由于间歇式送风机作用及防尘密封引起的干扰之危险。新型机器的制动由Benninger 与 Karl Mayer 研制。

Karl Maryer 还提供一种结合打结与切断新型的筒子架，Benninger 的架上有一个非常有效的吸风除尘系统和一个经丝交叉校正装置。

Karl Mayer 已经开发了一种适用于裂膜的滑离筒子架。这种机器有两项技术创新，它采用了控制预加张力调节器和压辊，压辊也起到了调节张力的作用。

整经机

滑动架控制的进步已趋完美。Benninger 已经生产出全自动整经机，而 Karl Mayer 的机器大体上是自动化的。单纱试样整经机已被进一步开发或升级。多色经纱的换筒已被 Suker-Muuer-Hacoba 全部自动化，而 Benniner 提供光控换筒辅助设备。

轴经整经机

轴经整经机方面从根本上没有新进展。现在有适合于 1400 mm 边盘直径的机型。最大运输速度为 1400 m/min。Benninger 提供自动化换筒。

浆纱机

浆纱技术维持不变状态,尽管如此,仍有一些值得注意的改进。采用变频控制的三相电机的单独驱动器、工艺数据的记录已变得更简单和广泛、比以往任何时候都有更多的公司提供液体消耗测定法,使机器看护更简单。

机器控制已被分散,每一个模块都有自身的存储程序控制器与中央计算机。Suker-Muller-Hacoba 继续是浆纱应用测定和控制系统的唯一供应商。

值得注意的是首次加到浆纱机上的软件系统——BEN-Size-Expert。它提供了适用机器定形和相关方法的浆纱应用的上浆纱基本数据。这个相关数据曾经也被引入分析与评估浆纱与织造性能。

靛蓝染色浆纱机

牛仔布市场的持续的增长,促使 Suker-Muller-Hacoba 和 Benniner-Zell 公司去研制新型机器。使用环保的靛蓝染色是一种趋势。这两家公司提供不同的方式,将亚硫酸盐的使用减到最小限度。

Suker-Muller-Hacoba 采用预先经过防腐处理的浓缩物,将浓度降到 2~3 g/l,避免不固色的危险,并用快速氧化法加速氧化(注入预热空气)。Benninger-Zell 染料在氮气气氛中染色与定形,从而形成无氧染色溶液。有了新的染整工艺控制设备的开发,Benninger-Zell 保证染色的一致性与染性,并贯穿整批织物的整个幅度。首次开发的是"BEN-Link"系统,适用于换批无间断的染色加工。

第19课 开口形成机构

在开口形成方面,新开发与更新的系统继续跟随着织物织造中更高的机器速度、更容易维护及灵活性的趋势。

织机上的引纬率已大大提高。开口机构的制造商则以一种更加紧凑的设计、更高的速度、更加稳定的新型机械作出反应。下面一些细节是值得特别指出的:

——开口形成机构与轴之间的传动部件现在已被集中润滑,而它们之间的连接已经加强。

——提花装置中滑轮机构不仅设计更大、更强,它们也用低维护钢球轴承来装备。

——提花机构从织机机身分离,具有独立的同步驱动。这防止了在长传动机构中的惯性危险。

——在高速电子仪器控制的提花机方面,从梭口上层经纱读数向梭口下层经纱读数已经发生了变化。借助这样的方法,通丝的运动更平滑,装置的使用寿命显著地延长。

——在狭幅织物织造中,织机速度通常特别高,在消极装置中的返回操作现在可以通过气流方式进行,因而有更大的灵活性。

——通丝回综弹簧的直径已制得更细,以降低抖动冲击的振动。

关于使织物生产更灵活,一些显著的改进被报道。提花机可以装入的竖钩最大数量增加了一倍或至少显著增加。一些整体系统的例子是:

——12288 竖钩（Staubli CX1060）

——10752 竖钩（Grosse EJP-2）

——13824 竖钩,含有 24 个提花机组件,每组 576 竖钩元件（Tis）

——5120 双开口升降机构元件,用于天鹅绒织造（Staubli 1090）

这意味着如果横跨全幅仅有一个循环,可以在织物宽大约 180 cm 范围,密度超过 60 根/cm。这允许织造任何要求的花纹组织,而不必考虑循环的连接,包括加边花纹组织,因为每根经纱能被单独控制。

当然,随着磁体数量的增加,能源消耗也越来越大,但在某些情况却大量减少。

提花机构里的丝线密度能用双目板修改,或借助于移动提花组件的附加装置的可调中间网格来修改。

电子提花装置的组件结构允许整排的组件放入或移出,以适合特别需要的竖钩数目。

在运转可靠性和织出织物的质量方面也取得了良好的进展。这包括电子提花机中的电磁监控,它甚至能与机器运转同时发生。机器密封装置防止了灰尘污染,在提花机上方,顶盖不需要额外空间,并允许强制通风。

这种新改进的控制器,没有任何需要。它们不再受到关于循环长度的限制。功能已被扩展,而操作已变得更加人性化。

通常与所有流行的计算机辅助设计系统也有兼容性。需要的地方,对花纹组织数据的修改,能当场在控制器上或在笔记本电脑上快速进行。现在,借助于手持扫描仪和笔记本电脑,制作简单的提花图案也是可能的。目前,一些织机含有鼠标和 IBM 键盘。

在厚重织物织造方面也有创新之处,最多适合 24 片综的电气自动控制的闭合开口多臂机可设计成为筘幅达到 3.5m 的宽度。

对于尚未准备好更换机械控制式提花机的织造厂家来说,一种电子读数系统可供韦多尔针机构使用。

传感器针靠压电方式被握住或者释放。在与电子数据处理系统连接的 13344-竖钩机器上改型是可能的。

第 20 课　织造的发展趋势(一)

在长期的停止不前后,织造领域有了再一次真正的革命——Sulzer Ruti "M8300"多相织机。该机在 1995 年国际纺织机械展览会上引起了轰动,基本原理是单相引纬,预示着几十年后织造技术的一个新时代。

自然产生的疑问是,这项新技术是否能成功,因为在以前曾经有过伴随多相梭口的引起的多种"幸福感",它最终失败了。这个新系统的前景似乎不太坏。在这样短的时间内谁会预料到这样高度开拓的织造系统被推出?然而,我们还是先来描述一下在传统织造生产中的最重要的进步。

织机的发展概况

——驱动、控制与计算机技术

最近几年,织机技术方面的真正革命源自驱动、控制与计算机技术。在驱动技术中,已经有了基本变化。用电子仪器控制的单个传动装置,提供了适合计算机控制的基础。频控电机或直流电机提供了大数量级的速度变换,适应纱线及所织造的织物。齿轮传动基本上是多余的。经轴与卷布辊由电子仪器单独控制。

目前,选纬、储纬器、多臂机和提花机都具有电子控制机构。用这种方式,在机器运转的同时,由计算机改变纬密和经纱张力、纬纱系统与织物组织选择性地计入是可能的。

先进的计算机控制程序允许重新启动,而不需要在所有的机织物上标明标记。完整的功能系列有助于以下所述:综片调平、停机时取消经纱张力、启动校正(张力)、在加速中超速、自动缺纬机构、自动断纬修复、切换到另一个储纬器、打纬控制及两端筘座传动。

所有这些用电子仪器控制的织机装置,都是从全电子计算机控制织机的基础上发展起来的,操作大大简化了。

——技术进步

新的传动与控制技术,也使科技进步成为可能。例如,受控的纬纱张力能使纬纱张力在引纬循环期间,与有关纱线及整个时间的变化相适应。这在所有的织造系统中(像片梭、剑杆和喷气织机)是有利的。就喷气织机来说,当纬纱突然停止时产生的高峰张力限制了的加速度,能被减少一半。这减少了在纬纱纱线上的应力,并提供了进一步提高喷气织机速度的机会。

电子控制单独传动的纬纱引出器系统,允许喂入装置可靠定位,已被导引的纬纱线立刻被送回到合适的运行位置,以防止在纱线通路中出现任何不必要的偏转点。

储纬器也得到了进一步改进。电子方法瞬时打开4个具有补偿作用的受控送经张力器。一个装有添加剂装置的储纬器装备用于加湿或向纬纱施加添加剂。可靠控制的织机后梁,一些尺寸大大减少,可以积极改变经纱张力的动态行程,例如,获得一个更快更清晰的开口高度。这意味着在经纱上的应力更低、平均经纱张力更低,速度更高。在开口大小的调整方面是电子仪器还没有渗入的一个区域。正确的开口配置,对织机性能的稳定至关重要。可惜只有少数织机,例如 Sulzer Ruti 制的 G6200 剑杆织机或 Nuovo Pignone 制的 FAST 型剑杆织机,为织造厂家提供了精确可调整的开口尺寸的装置。

第21课 织造的发展趋势(二)

传统织机中的改进

——片梭织机

与全电子控制织机的发展趋势相矛盾,并装备了各种各样经过改进的装置,这自然是不便宜的,简单的配备标准织机的开发,成本较低。一个例子是 Sulzer Ruti 产的 Plean 型片梭织机。在片梭织机方面,引纬速度大约为 1400 m/min,甚至高达 1500 m/min 筘幅为 3.92 m。片梭织机在特殊宽度方面可达 8.5 m。

——剑杆织机

剑杆织机仍有发展前景。例如,剑杆头的设计正在大为改进。更小更轻的剑杆头允许更高的速度和更低的开口高度。这意味着经纱上的应力更低。剑杆织机上的入纬率确实已达 1000 m/min。"漂移"的剑杆其不需要任何伸入梭口的导件，在梭口的开口方面具有优势，对经纱末端的损伤较小。

Picanol 公司现在已把喷气织机的快速换批概念应用到 GTX 剑杆织机上。像在"Omni"和"Delta"类型中的那样，全部经纱组件现在也能在 GTX 上变换。

ICBT，众所周知的变形机械的制造商，已经从 Vamatex 公司接管了原来的 Saurer-Dierichs 机器，并提供两台有受控剑杆的机器和 Picanol 的情况一样，这些机器装备有经纱组件，该组件也是自动移动的，换句话说，它负责输送。从成本方面考虑，这应该是有吸引力的。ICBT 引进了一种新型提花系统，在该系统中提花机架构成织机机架的一个集成部分。结果是一种惊人的低结构就振动而言具有优势。

Nuovo Pigpone 有其代表建立在 FAST 系列基础上的一种典型的 NEXT 织机，其具有未来主义风格的涂层，旨在对降低噪音作出较大的贡献。Somet 也提供包含降低噪音的机器涂层，据称减少噪音 2dB(A)。另外也有一种新型联合的钳纬器与剪纬刀，在没有折边的帮边下，其在送纬侧减少了纬纱的浪费。另一家制造商苏尔寿·吕蒂也展示了一种减少纬纱浪费的特殊的钳纬器。

在 G6200 上，它在左侧纬纱的损失减少到 4 cm。任何使用纬纱的人，都能马上意识到这种潜在的回报。

Sulzer Ruti 研制了一种毛圈织机上的电动毛圈高度控制机构，它展现了新颖的设计理念。

剑杆织机断纬的自动化修理费用相对较高，仅有 Somet 和 Vamatex 公司提供。Somet 采用了一种机械原理，而 Vamatex 应用了一个气压系统。

剑杆织机的制造商们，用许多采用最细和最粗纱线的特殊织物，展示他们机器高度通用性。例如，Dornier 推出了"在移动中"的设计变化，它实现了在剑杆织机上不用停机而能改变风格。当然使用同一种经纱，但引入四种新纬纱，而纬密、速度、经纱张力及组织发生了变化。

根据剑杆引入原理运行的地毯织机也必须提及。这里，Van de Wiele 研制了第一台有三个剑杆的刚性剑杆织机。靠同时引入三根纬纱，在相同速度下，实现了 50%的更高生产率。增加速度也是可能的。

——喷气织机

喷气织机的制造商们提出证据，即他们现在不仅能提高生产效率，而且也能织出难缠的纱线及厚重的织物。高捻绉纱线、绳绒纱、娇嫩羊毛及长丝纬纱正在高速织造，还有在高生产率状况下的牛仔布及在超过 1000 转/分速度下的提花图案。

Dornier 演示了有绳绒线布边的毛圈织物的织造，但在这种情况下速度不得不暂时降低以适合引入绳绒纬纱。

Gunne，是几年前首家提供毛圈织物喷气织机的制造商，它推出了一种有低振动机架的新型频率控制机器。

Picanol 在喷气织机上展现了高灵活性。例如，在一台 Omini 机上一种高捻绉纱在新型"延伸器"喷嘴的帮助下被引入。一种绳绒纬纱的装饰织物，以 750 转/分的车速织造。Omni 是为灵活性而设计的，而低成本的 Delta 机是为在更合理的成本条件下，特别为生产标准织物而设计的。Picanol 也已研制了一种纬纱超过正常范围传感器。

第 22 课　织造的发展趋势（三）

值得一提的是 Somet 制的双幅 MACH-3 喷气织机。这是一种具有单驱动的混合机器。由于经纱通道的垂直安排——经轴在下与卷布辊在上——双幅机器占有的空间与单幅机所占空间一样。每幅宽度为 3.22 m，运转速度为 600 m/min，入纬率为 3864 m/min，是 1995 年国际纺织机械展览会上，引纬速度最高的喷气织机。

Tsudakoma 开发了两种型号的机器。ZAX 是打算用于以最大速度生产标准织物。目前，一台 1.7 m 筘幅织造棉织品的这种型号的机器正以 1700 m/min 运转，而 2890 m/min 的入纬率在单幅织机上是最高的。但 ZA209i 的实力在于它的灵活性。用于毛圈织物的，有 ZA207i。AFR 通用全自动断纬修复系统，也能修复主喷嘴之前的断纬。

——喷水织机

喷水织机在欧洲很少使用。在力图织造不上浆的长丝经纱方面，应该认真考虑使用这种织机。随着接力喷嘴或纬纱控制掣子的不存在，这些机器用于织造无浆空气变形长丝纱是理想的。

——多臂机与提花机

喷气织机所追求的灵活性，对多臂机和提花机制造商提出了很高的要求。要求生产出能够高速运转的相应高效的设备。电子多臂机如果只用几片综的话，速度可达到大约1000转/分。用16片综时，可获得的速度约为750转/分。凸轮开口机器可达到1100转/分。当使用20或更多综片时，回转多臂机的转速为500～600转/分或低一些。

  Tsudakoma开发了一项有趣但尚未全面大批生产的创新，既每片综单独驱动。靠这种方法，在每片综上开口闭合运动周期及时间可以独立控制。Grob制的新型Optifil综片值得一提，它允许纱线更平滑地通过。

  提花机的情况，尤其是综丝包括弹簧回返零件的情况下，特别不容易。近年，由Bonas和Staubli研制的高速提花机，以超过1400转/分的速度运转。实际上，人们可能认为这样高的速度和需要的平滑运行在目前将是难以达到的，因为就在几年以前，当电子提花机被引入时，每400转/分那样低的速度被认为是一个巨大的成就。在有单丝控制的仍然缓慢运转的整体式提花机上的改进，Grosse与Staubli推出了具有多于10000竖钩的提花机——由于通丝磨损问题仍被加剧，因为在竖钩上的载荷，必须由单个通丝及其回综弹簧通过增加的张力来施加。

  提花机独立驱动的试验已经开始。

  Suhleicher用一个双金屑吸盘读入竖钩，换句话说没有磁性。该系统能源消耗低，因此产生的热量较少。到现在时间为止，该系统仅作为用于针栅格替换部分。一种柔性提花机，根据单动原理操作，没有滑轮升降系统、采用标准设计，由Tis研制。每个组件最多控制铝合金576根经纱并滑动安装在主机架上。一块特殊的目板能使经纱密度变化。单独组件的排顺能使通所运动大体上是垂直的。

  ——织机环境

  在织造过程中，环境条件尤其是相对湿度，对织造工程影响很大。具有单独空气调节的织机，在除尘改进方面有了最新的进展。对于短纤维经纱，有一个在织机后梭口区域有强迫向下的空气流系统，靠这个向下的经过调节的空气流，促使飞花在地面上沉积是个优点。飞花即从工作区域向下带走。带尘空气的排出位置最好设在后部梭口的下面，因为这是纤维磨损最多的地方。像LTG和Luwa提供的这种系统，改善了工作区域的条件并产生更好的运转性能。直接的机器空调也是最经济的系统。

第23课 织造的发展趋势（四）

  ——品种变换与经纱变换

  在近年，自动卷布辊系统成为大多数织机制造商的发展方向。在这一水平上

YARN AND FABRIC FORMING
现代纺织英语

的自动化,事实上是最先进的。但必须始终与是否具有大匹长度与适合织物检验的机会的争论相抗衡。

经纱变换与品种变换,从经济的角度看是特别引人注意的。如今,快速换批(QSC)实际上所有的织机制造商都提供,大多数采用 Genkinger 公司的快速换批小车。这种小车现在能自动调整与织机平行,使快速加载到机器成为可能。有一种新型快速换批小车,具有适于同时装载两个完整经轴的装置。随着快速安装综片结合件的开发,装卸也被进一步改进。

Picanol 快速换批系统不同于其他制造商,它有一个能从织机上拆卸的组件。这使经纱准备能在穿经工段基本上完成。Picanol 还开发了一个建立在风格变换系统上的快速换批系统,而结经在穿经工段完成。

多相织造系统

在 Sulzer Ruti 多相开口系统中,若干平行的开口波形成,它们沿着经纱方向移动。M8300 采用了"织造的旋转件",它有一些开口形成系统。经纱围绕着旋转件部分的弯曲。因而连续形成四个开口因而是可能的。由于旋转件的直径小,所以开口的高度非常低并且摩擦接触区短。开口由导纱梳栉来完成,它借助于短的侧面运动判断是否经纱被放在或靠近开口形成元件。开口形成的气流包括一个空气通道,经过它意味着纬纱被引入。利用压力非常低的喷嘴,纬纱同步地连续引入四个开口。M8300 配备多台微处理器,具有全电子控制功能。它有自动选纬和自动断纬修复的功能。该机加工达到 1.6 m 直径范围内的经轴。它已装备了快速换批系统能更换安装在机顶的卷布辊。

M8300 已经处于开发的高级阶段。与 20 世纪 70 年代的机器不同,该机实际上能代表织造技术的新纪元。首先,它已具有极高的性能,即每分钟超过 5000 m 的引纬速度加上相当大的发展潜力,其次,这项新开发由强大的织机制造商支持,再一次与 70 年代的发展形成对比。在这样一个复杂的新开发中这是最重要的,因为传统织造技术几乎不能引进任何东西,但所有重要的功能元件都必须从基础设计出发。这项新型织造技术的前景,可以说是好的。

引纬速度

曲线图(图1)说明了国际纺织机械展览会展示的各种最大引纬速度。剑杆织机毫无例外地超越 1000 m/min 的记录。片梭织机现在达到 1500 m/min。单幅喷气织机的引纬速度为 2890 m/min,而 Somet 双幅喷气织机引纬速度为 3864 m/min。如果我们追溯到 30 年前,在传统的单相织造方面,很明显已经取得了迅速的进步。

这个尖峰,引纬速度以具有 5500 m/min 的多相织机性能达到顶峰,标志着织造技术新时代的来临。

第 24 课 生产地毯的机器

借助于计算机技术,地毯织机的联网已得到进一步完善。这包括较高的地毯质量及生产率、较短的换批时间等。

簇绒机

TX300 簇绒机由 Tuflex 公司推出。微处理器直接控制应用伺服电机的机器功能。这使物质连续不断向下移动与高生产速度变为可能的。这种簇绒机包含两个滑动式针座用于交错绒头,每一个有伺服电机控制。被生产的花纹靠由计算机控制的伺服电机可任意编程。该机的主要特点是,绒纱以伸直的状态被喂入,适于线型或交错绒头的精确绒纱喂给。

剪刀状夹片和刀片,已被剪切机型的刀片运动而取代。这些特点是为了消除用于割绒地毯的剪毛,因此有效地节约了绒头纱线。

地毯织机

现代织造设备似乎要以电子花型控制的双层绒头剑杆织机为特色,采用电子花型控制,在由可获得的绒头纱线决定的颜色调色板内提供快速花样变换。变换花型与单循环、双循环或三循环操作间无问题的绒头组织更换有关。当改变链经纱和衬垫经纱的织造根数时,可在短短几个小时内改变综片凸轮。借助计算机辅助编织,可以通过按筘幅与长度来配合批量以减少切断的浪费。这是 Van de Wiele 和 Schonherr Chemnitzer Webmaschinenbau GmbH 公司的机器的标准设备。

与此同时,Schonherr 开发了双循环双纬起绒组织,不受在花样与颜色方面模糊限制的影响,Ven de wiele 推出的一项革新是,在三剑杆织机上进行传统的三纬两循环织造。用这种织造方式,自 1932 年起它以三循环的形式被广泛使用,织造地毯时,地毯背面花较清晰。埋头绒头均匀分布在在上层和下层地毯之间。

一种新设计的电子控制的复动式全开口提花机 CX1090B 由 Staubli 研制。8 色绒头适于形成花纹。利用完全令人印象深刻的、电子形成花纹系统的多功能性,在织物上织出 36 种不同色系,包括混合色。

Hemaks 公司提供一种双层绒头地毯织机,有非常简单设计的单挠性剑杆,适合于有摩擦背面的地毯生产；另一种有两个挠性剑杆，用于传统的三纬起绒组织的地毯机。Verdol 提花机连同通丝和背面拉耙机构都由该公司制造。这种机器在土尔其及中东主要由小规模生产者使用。

为减少机器的清洁时间，地毯织机的制造商应该考虑提供附加除尘系统，适于用在引纬和/或综片区域的可能性。当织造黄麻和/或羊毛纱线时，这些都是有利的。

地毯提花机

在提花工程方面，一些供应商的一项创新，用特殊陶瓷压电元件取代用于电子花纹控制的磁件。这些（元件）起弯曲谐振器的作用。压电元件具有以下优点：防尘密封盒或组件的尺寸小，功耗低，不需要冷却。两个半竖钩用一个全竖钩代替。Schleicher 和 Takemura 为地毯织造和单层织造的所有机械控制的提花机提供压电元件的改型装置。

它们取代了 Verdol 和 French 细针距提花机纹板。与电磁控制相比，压电元件能使能源消耗 95% 以上。Takemura 通过光波传递花纹信息。适于 28 针的 Verdol 的一个压电组件仅 5 mm 厚，只需要 50 W 来驱动 1344 竖钩。

Karl Mayer 公司是提供适合其彩色提花窗帘布机器压件基础零通丝提花机的首家公司。不用的供应商已升级了他们的磁控基础系统。

各种地毯织机

Karl Mayer 推出一种地毯针织机，适用于提花图案的起圈绒头地毯，它以每分钟 250 线圈横列的速度生产。形成花纹绒头的纬纱线，通过经纱的编链组织被引入。四根埋头绒头，在形成花纹绒头的毛圈处呈延伸形式。纬纱线覆盖地毯的背面。这种有 5 或 8 绒头毛圈/英时的地毯，在订货方面取得成功。

Cotinfi 提供一种机器适于生产有粘合切断或采用一种升级的纱线折入技术（无折刀）的毛圈地毯。采用非织造布黏合，生产起圈绒头地毯是一项新技术。随着填充材料的掺入，PVC 黏合剂的用量已从超过 1 kg/m^2 降至 0.8 kg/m^2。

据称这种粘合剂在绒头纤维之间具有更大的渗透性，较少地渗入黄麻背面。后期能从 407 g/m^2 降低到 270 g/m^2。

第25课　试验与测量仪器（一）

纺织原料试验

显然，纺织机械制造商非常重视棉纤维准备中异纤维的检验，并开发了一些系统。著名试验设备制造商 Zellweger 着手解决此问题，并推出 Optiscan 异纤维检测系统，作为一种设备用于纺纱准备。

长期以来，原棉中的杂质如聚丙烯包装材料的碎片、细绳、有色结块物质或织物碎片加上纱线、织物及成品织物中的综合疵点，一直是造成费用高昂的原因。

——纤维杂质测试仪

1995 年，在米兰由 Maschinefibrik Rieter 推出的作为一种离线测试仪适于棉中的虫污或黏着性的"纤维杂质测试仪"，引起试验设备的制造商轻微的轰动。在以色列由 Dr. Uzi Mor 研制的该装置，实际上回溯到 1994 年，当时其发明者在不来梅发表过一篇论文，该论文对此作于描述。

由于目前尚没有成功操作用于测量"虫污"的装置（NIR 测量到目前为止没有得到承认），所以它能被无可非议地称为一种创新。这个装置还可以对毛粒和种籽壳片进行快速测试。

在近期，虫污问题伤害棉花产地已比几年前更厉害了。一种有效运行的可靠装置可用于黏着性的早期检验，在原产棉国，早在轧棉阶段，或至少在纺纱准备阶段运行将是非常有用的。但是，一个不可忽视的问题是，对时有时无发生的虫污的代表性取样。

——成包原棉的测试

HVI（高容量仪器）棉测试线已获得世界范围的承认，适用于成包原棉的测试。根据最近的 ITMF 报告，超过 1200 条线正在全世界范围运转。在汉诺威举办国际纺织机械展览会时，有大约 500 条线，其中差不多一半在美国。

对 HVI 线使用的强有力的促进因素，不仅来自于世界各地的纺纱者或适于质量最优化或混合组成的在线编程系统，也来自于纺纱机械制造商（尤其来自于 Schlahorst），而且还来自于棉花生产国家。在美国，HVI 测量形成正式用于陆地棉的分级基础。从长远的观点来看，其他主要棉花生产国必定会效仿。

由于 Spinlab 和 Motion Control/Peyer 集团现在已被 Zellweger Uster 接管，目前世界上仅有一家 HVI 线的制造商。最新的 HVI 线能由一个人操作，稍加修改，还提供了适用化学短纤维的质量评定装置。

纺纱准备阶段的生产控制与优化

一种合适的生产控制工具，特别适于纺纱准备中的最优化梳理，是 AFIS 系统适合快速非主观地测量棉结。一种新产品是快速单纤维成熟度测试仪，根据 AFIS

系统发明者的原创思想研制。

经过长期试验和证明的ⅡC-锡莱纤维细度/成熟度测试仪已经过了进一步的开发，并在米兰以 Micromat 高速测试仪的形式展出。有加速试验循环的装置能与现有的 HVI 线结合。

一种同样证明是成功用于转杯纺纱最优化的装置，尤其在原料选择方面，是 Quickspin 系统，它由邓肯道夫纺织工艺技术研究所研制。这作为 MDTA 系统销售的。除了灰尘、杂质和纤维碎片测量之外，还可以预测纺纱性能和预期的纱线质量。混纺优化也是可能的，不仅适用于三罗拉纺纱，而且甚至适用于粗梳和半精梳毛纺（至少达到 60 mm 纤维长度内）。如 Reutlingen 试验所证明的那样，短纤维亚麻组分的质量评定是一个进一步的选择。

第26课　试验与测量仪器（二）

一系列"低成本"纺织设备吸引了很多人的兴趣，尤其是不愿意或不能立即投资 HVI 线的纺织厂家。这些是建立在国际上成功的由原来的 Spinlab 公司制的以前的装置（730 纤维长度照影仪、750 色泽仪、775 马克隆尼气流式纤维细度测试仪、380 纤维紫外线荧光测试仪）基础上的。尤其应该是最后提到的（那种），具有感色灵敏的棉批组合（用于天鹅绒！）甚至用于评估化纤上的添加剂。

化纤也能用一个改良的"AFIS"系统来测试。因为 2000~3000 根纤维仅在几分钟内能被测定。用比传统的单纤维测试更好的统计置信度，可以快速的得到原料细度和长度分布的结果。

对于传统的化学纤维的单纤维测量，仍然按 BISFA 标准分级，引进一种新的仪器是 Vibroskop 400。在测量准备中，单纤维只需要附属的夹紧装置并挂在框架中，接下来试验自动进行。

作为一种对牵伸机构变量校正控制，条子均匀度的在线控制，早已变成著名牵伸机构制造商的主要卖点。控制系统提供进一步的辅助装置，一些主要机械制造商如 Sliver master（Loepfe）、Sliver Dance（Barco）或 Sliver Control（Zellweger），等等。集中的数据收集在这里也提供了机会适于及时的生产与质量控制。一些监测和均匀控制系统是由机械制造商开发的，直接安装在各种开清棉机的喂入机构上，越来越趋向于简易操作。

纱线制造与纱线测试

——短纤纱

在转杯纺织机上，常常需要对单独纺纱头进行监测，促使提高纱线的质量，以适合下游的加工。环锭细纱机仍然存在问题，尤其在连接系统，尽管最初的在环锭细纱机上尝试纺纱头处监测已有相当长的一些时间。更早期的解决办法大多数失败在成本的控制上。

1995 年 Maschinenfabrik Zinser 曾与 Schlafhorst、Loepfe 及瑞士纺纱者 H. Buhler 合作展示一种新纺纱头识别系统。在该系统中单独的管纱支持器安装有一个任意可编程的集成电路块，每个芯片都结合了纺丝头的质量与产量数据以及络筒机头的质量数据。它还能识别有缺陷的纺纱头，并根据预先规定的公差限制分类出不良管纱——这是朝着最佳纱线生产方向迈出的又一步。

在米兰展出的一种方法，它立刻吸引眼球，模仿机织和针织织物并预测在成品织物中可能的疵点，是高度设计的新 CYROS 系统，由 Reutlingen 的 Zellweger 推出。该系统使用由 G580 光学纱线均匀率测试仪提供的真实数据。CIS 公司在机织织物方面的设计经验及(美)棉花公司在棉花和其加工过程中带来了令人惊异的现实。在络筒机头或纺纱杯头采用同样的方法，在不久的将来，在那里光传感器也可以使用。

——长丝纱线

杜邦公司开发了一种新的仪器，用于测试长丝纱线。自动化的纱线细度测试、拉伸力测试和交缠测试加上关于 POY、MOY 和 LOY 纱线及变形纱在热效应下的自动收缩力的测量，由传统的长丝纱线试验仪器制造商提供。

这些制造商再次集中注意力在他们的长处上，并采用了值得注意的改进。例如，Dynafil M 装置适用于拉伸力试验与收缩力测量，用自动喂入变换器(悬挂达到 20 支卷装)来装备。在捻接器的帮助下，卷装在 20 秒钟内变换。

在变形方面，在变形头处的丝线张力监测，采用在线质量控制的形式正变得越来越多。

在织造车间成功的质量汇总，这里不详细讨论，但 Tensojet 断裂强力测试仪值得一提。该仪器每小时在纱线上执行达到 30000 次断裂试验，所以在一定程度上，在卷装内随机低强部位的检测成为可能，这在以前是不可能的。

第27课　针织横机(一)

总的发展

对针织横机机械不同的制造商，值得注意的是在早期国际纺织机械展览会上，作为创新而推出的一些成果，现在已变成工艺流成。在针织横机方面的改进，正变得越来越趋向合理化发展，使针织生产更灵活。所谓的"快速反应"，即对市场需求作出更快地反应和以高质量标准进行各种的设计，越来越小的批量化快速生产。有时候，一些机器展示出全成形设计的广泛多功能性。现在，大部分针织横机安装有全成形装置作为标准装备。

这些包括一个在针床下面的二级织物卷取机构、沉降片、丝线长度控制机构以及在紧凑机器上适合于在空针上针织开始的脱圈沉降片。长度控制首先由 Shima Seiki(DOCS)在 1985 年大阪国际纺织机械展览会上发表，它现在也被 Stoll(STIXX)、Universal(始于 1996 年中期)和 Protti 用于保持针织嵌条的尺寸不变。该原理是建立在一个针织线圈横列纱线长度的测量与校正基础上的。纱线经过一个测量轮，它每转一次就能产生一定数量的脉冲。在操纵台上的计算机根据线圈横列中监测的目标长度变化，通过调整脱圈三角保持针织嵌条的长度恒定。

辅助针床的广泛应用

辅助针床的广泛应用(它被安置在针床上面并装有移圈针)，清楚地显示了针织横机制造商特别努力使全成形针织和一般针织工艺合理化的努力。这种技术曾在米兰有四家机器制造商展出。Shima Seiki 和 APM 提供有两个辅助针床的紧凑型机器，同时 Rimach 和 Comet 的机器有一个辅助针床。

就 Rimach 和 Comet 来说，第三个针床是以这样的方式被分配，即当移圈发生时，这两个辅助针床同时向外移动或向内移动。当收针时，右手或左手针织边在一个三角滑座行程内因而被同时向内移动。除了更高的生产率之外，有附加的成形花纹装置，正如双面针织物在辅助针床上被"放置"一样。

新机器开发——整体针织套衫的编织

回溯到 1991 年该项技术曾在汉诺威被展出，尽管以开放的形式("集成"的编织)。1995 年在米兰，首次展示了零缝制的针织套衫。就 Shima Seiki 来说，整体的针织套衫在两种新推出的机器上织出，四针床 SWG.X 和双针床机器 SWG.V。在 SWG.X 上编织一件运动衫所需时间为 35 分钟。

全成形或三维编织，在产业用纺织品领域也特别有利。与其他织物生产方法相比，平型编织技术不仅提供零接缝产品，而且也具有高度灵活性和产品的多样性。在厚服装条件下，把丝线在力流线的方向排列是可能的。Universal 与一家汽车制造商合作，推出了一件非常有趣的来自于汽车方面的样品——一件整体的汽车座罩(座、扶手、头靠)，是在 Power-Pressjack 的 MC-748 型机上编织的，带有电源插孔。反映了针织物在技术市场的重要性。

第28课 针织横机(二)

无三角座滑架的针织横机

近年来的创新中,Tsudakoma 制的 TFK 无三角座滑架针织横机引起人们特别关注。该机在灵活性、生产效率和生产可靠性方面提供了一些新的装置。机器的主要特点是单针控制,每根织针由其自身线性电机所驱动。每根针可独立地任意控制。在一个线圈横列,从总的 80 种可能的线圈长度针织最多 30 种款式是可能的。它无疑为新设计结构提供了广泛基础。这些优点是令人印象深刻的,尤其是在机器生产效率的潜力方面。具有 122 cm 宽的针床和 15 cm 的退圈高度,TFK 能与八系统针织横机相比。随着三角座滑架往复运动引起的时间浪费被完全消除。在线圈转移方面,在一个线圈横列的若干区域能被同时转移。这些机器在可靠性方面还提供新的技术装置。

机顶式直接喂入减少了总的纱线包角,使更低更均匀的丝线张力成为可能。在织针控制方面,针织最大的平滑性,有 250 种三角变化选择。这样,就有可能客观地考虑不同线圈结构或原料的具体编织要求。例如,它特别适合针织低伸张或脆弱的纱线。最后要提到的是机器简单的维护。轴向针驱动避免了剪切力(过去常在针道中产生),这将会减少机械磨损。除了提到的积极方面,关于设计的通用性、针织可靠性和机器的工作能力,许多问题仍有待解答。这些只能通过工业实践来回答。

——高生产率

一个显著的特点是机器制造商正作出巨大的努力,进一步向更高生产力和更多灵活性方面扩展现有的平型针织技术。在此范围,Stoll 推出了其"CSM340.6"四系统紧凑型机器,而 Universal 推出其"MC-888",第一台八系统针织横机。随着 Universal 的 MC-800 系列的问世,新的领域被开发。机架结构紧凑符合人类工程学方面设计,新型的标准组件三角座滑架控制机构,形成三角座滑架的集成部分,承担针织需要的控制功能,同时车载计算机负责数据的记录与读出。

这些优点包括车载计算机工作量小、具有更简单的牵拉辊结构、数据传输可靠性高。stoll 介绍了一项有趣的细节改进,优化了嵌花线控制,借助它最大速度已从 0.7 m/s 增加到 1.0 m/s。机号自 1991 年国际纺织机械展览会已被增加到

YARN AND FABRIC FORMING

现代纺织英语

E14，而在 Shima Seiki 的场合为 E18。

更大的灵活性与机器功能的扩展趋势无间断的持续着。例如，Shima Seiki 的 Split Stitch Function（机器划分）目前是最新的。上述情况也适用于对线圈纵向脱散的抵制，在控制方面已被主要机器制造商们以不同的方式处理。针织横机与设计工作室联网，现在由 Stoll 和全球公司获得。对于已经有这些装备的针织厂，它是生产监测和检验分析的重要工具。

设计系统

正如机器那样，设计系统的研发部集中在快速反应程序上。程序正变得越来越容易使用。自动化组件促进了编程及减少在制作控制程序中所需的努力。在这一方面已经作出了巨大的努力，尤其在全成形针织领域。组织结构的仿真目前是最先进的，某些机器制造商，可以选择它。在计算机系统方面，遵循两条路线。对于 PC 基础的系统，首先考虑的是灵活的低成本计算机辅助设备。

这些系统常常用 Windows 运转，使 Windows 程序在一定范围得以使用，就建立在工作站基础上的设计系统的针织横机制造商来说，系统的生产能力有集中优先权。这就是如何满足日益更加复杂的程序的增长需求。

第29课　大直径圆机及辅助设备（一）

圆形针织机的灵活性大

今天，圆形针织机的织造工艺越来越具有通用性。在保持高质量标准的同时，高灵活性与短的更换时间及合理的（经济上可行的）生产结合，特别有利小批量（布匹与成形嵌料两方面）的生产。灵活性对机器制造商意味着，能马上用结合的机器生产出适合由当前市场需要的产品。灵活性也意味着即使只有一套设备，适于在相当短的时间内将机器转化成有关产品的生产。这不是什么新鲜事，过去许多不同的解决办法已经被提出，但需求随着短期时尚新的市场渠道而变得更加严格。

制造商们按照"快速换批"的概念来概括制造不同产品的高度灵活性。针织生产者们目前涉及的概念有：

——改变机号（织针隔距）

——改变针筒和针盘的直径

——机器类型之间的组件的兼容性

——编织组织的快速变换

——更快的式样变换(例如,彩色提花或调线选择)

——高标准的用户友好与简单维护

——机号与直径的改变

"快速变换"方法对一些在修改机架方面的机械制造商(例如 Orizio、Mayer&Cie、Monarch、Pilotelli、Terrot)已经是明显的。就大直径圆形针织机来说,机架通常是标准三脚门架加上在计件衣坯机中的一些支架类型,可以轻而易举地从 30 kg 或 50 kg 布卷,转变成 120 kg 布卷。机架被限制在只少许几个使用不同尺寸的针筒与针盘可以改变直径。

两个动力输线支柱或输线轮导纱器圆环支架之间的距离,大到足以能使针盘与针筒被水平拆卸。近年,"快速变换"主要与没有单针选针装量的平针织物机器有关,换句话说,与平纹针织机有关。变换三角部分,例如,改变结构,从三线起绒变成单绒头织物或平纹织物。在喂入组内转变机号无需变换织针三角和导纱器。

改变机号所花的时间,在平针织物机器上没几个小时,同时对双面针织物机器需要 1.5~2 天。改变机号装置相当于机器成本价格的大约 20%~25%,而直径变化为 30%~40%。

——无单针选针系统的针织结构

目前,客户能在低灵活性的高速机器之间进行选择,用于平纹织物、罗纹或双罗纹组织(需要封闭针道),这些机器是为标准织物或用于印花织物而建立,更灵活的机器能够覆盖广泛的针织结构。一些高速机器正以 2 m/s 的圆周速度运转,由于它们众所周知的高噪音而令人不快。目前 1.5 m/s 的圆周速度已经成为标准。从双罗纹组织改变设计装置,例如有变换三角的机器现在是最先进的。这些机器非常适合生产布匹,如印花织物。在需要频繁变换花样的机器上,采用附带变换装置的三角可以减少机器改装时间。

一种包括三路技术的八三角机器引起业内人士的关注,其三角可以在缓慢模式下加入转换成另一种设定(RDS 压针三角系统)。(美)杜邦公司已经研制出一种新罗纹组织结构,在每个线圈横列加入一根弹性丝线。每一路进线中均有底线(棉)被编入。仅有罗纹针盘针作弹性添纱,下一路进线中则有针筒针完成添纱。一种需要改进的三角结构,由 Terrot 推出,该技术生产出一种高弹力针织物。

YARN AND FABRIC FORMING
现代纺织英语

——具有电子单针选针的提花机

针织者面临着各种各样的新型和升级的机器,这些机器都有电子控制单针选择。用户可在有明显成本优势的、无限循环规模的、由电子控制的单针选针及双路技术或更昂贵的三路技术之间作出选择。

第30课 大直径圆机及辅助设备(2)
——值得注意的发展

由众多制造商制造的双面针织机,其设计的多功能性扩展了条纹或移圈装置。

一台双面针织机,装备了有握持/脱圈沉降片圆环。这使得设计可以采用纵向的条纹在针盘中使用织针编织带。像平针织机那样,这对脱圈之后重新开始编织有帮助。

单面针织物机器与4、5、6色调线装置结合,由许多制造商提供。有快速安装装置的调线装置使运转中的简化成为可能。

适合于生产外衣织物和汽车、室内装潢座套的长毛绒机器种类繁多,其中有一台备受推崇的彩色提花全长毛绒机器,用其生产大花纹图案,有12种那样多或最多16种色是可能的。借助于使用特殊的沉降片,长浮线及相同长度的长毛绒是可能产生的。

针织纬编的高性能圆型毛衫针织机,是针织横机的主要竞争对手。具有旋转针道的机器(如 Jnmberca 产的 TLJ-6E)则显示出巨大的发展潜力。在有电子单针选针的圆型针织机之中,它们通过针筒内三路技术在两个方向上的线圈转移,5色调节装置,由正负5针和握持沉降片使机器实现动作变换及开锁器架起,具有最高的灵活性标准和通用的设计潜力。同样有趣的装置,如曲线形调线装置这样的机构,当不使用机器的整个圆周时,减少纱线消耗。

嵌花花纹到目前为止,只能在针织横机上进行。(德)迈耶·尔公司研制一种嵌花双面大直径圆形针织机,该机每分钟能产生40次往复运动或线圈横列,进程量达20次。用这种机器,以比针织横机大3~4倍的速度,生产对称嵌花花纹图案。

低剪裁无边缝内衣,现在能在有贴边的针织半成品嵌条装置的、专门化平针织物机上生产。

在细罗纹织物的针织方面，有比较广泛的形成花纹装置，在能生产高质量的细罗纹提花与2∶2罗纹交替使用，不降低针织品质量。能广泛提供带累纹滚边的移圈针织机。

值得注意的是，并非所有的机器制造商能成功地将电子形成花纹纳入机器在车脚中。有时候，大型控制组件妨碍访问与观察。并入机架中的电子系统提供了一种示范性的解决办法。

著名部件供应商的电磁选针器系统，被应用在许多的机器中。通过一个或两个针踵来进行选针。

不同机器制造商的设计程序，能在CAD系统编译，也能直接在机器上进行。目前，几乎没有完整的设计系统由机器制造商提供，而是适合于PC的软件。

机械单针选针的提花机

许多单面针织物与双面针织物机器装配了机械提花选针，作为全电子选针的替换物；在这种情况，提花宽度一般为36针或72针。双路或三路技术的小型提花选针系统，和提花轮、提花滚筒、提花圆盘及提花梳片，可从主要的或许多新的较小的供应商那里可获取。

该技术已经证明，对于特殊产品如花式组织结构、小型提花图案等方面是成功的，它是简单与坚固的，有积极式选针，相当便宜，管理上没有任何昂贵的基础结构如CAD等，有时候还允许更高的圆周速度。

第31课　大直径圆机及辅助设备（三）

多种发展趋势

目前，有一种明显的适合于外观粗面的粗针距机器的发展趋势。

可编程的三相频控控制传动，今天成为最先进的，可以精确地确定全速和慢行时的平稳运行和最佳速度。一种用于圆型针织机起动/停的手持无绳手控器已推出。

沉降轮已经被许多制造商移到外喂纱区，这有助于转化任务。用容量50 kg的割绒装置替换织物卷取辊，可在15分钟内完成。

Groz-Beckert公司研制了一种"曲折""低壁"织针，它的凹槽里装满塑料。这样防止工业灰尘在针槽内堆积，而不影响针的已经知道的益处。优化的针织织针也被推出，它具有锥形针钩以减少在编织中织针变形的危险。

YARN AND FABRIC FORMING

现代纺织英语

在沉降片上和导纱针上更好地修整边缘,加上一个高标准的磨损防护,有助于提高针织元件的寿命,并使织物更加一致。

不同制造商都以他们的最小化区域、光滑机架、明亮的工作区域及容易接近的针织区域而著称。

机器模仿有更大的用途,它适合于说明特殊运转状态或保养周期,显示用于织物卷取需要齿轮配合。

飞花吹拭器、油缸、速度等,按照可编程中央控制系统的要求进行监测和调节。许多供应商正在试图靠低成本系统来降低机器的成本,比如齿轮或V型皮带驱动的织物卷取。几乎所有的针织机制造商,都为他们提供有铰链的织物牵拉和卷取装置的机器。

自1991年国际纺织机械展览会起,一种加剧的趋势是依靠机架的标准化来降低机器的成本。

适用于圆形针织机的辅助设备

新颖的"混合式"喂纱器被推出,可用于常规的(正喂)或间歇的(提花)纱线消耗。

新型的弹性喂纱器现在被更容易装载,而筒重高达一千克,并能更好的防止飞花沉积。

注意把兴趣加入到一种带熔接装置,有增加管状导纱器的筒子架的用途,它使喂给传动带在机器上直接被熔接在一起成为可能。

提供了进一步的质量保证机会,通过纱线喂入长度的连续控制。通过传感器罗拉监测规定的喂入长度,并与允许极限进行比较。如果超过允许限度,机器停车。

常见的鼓风机系统(巡回式飞花吹除器和压缩空气喷嘴等),已被"振动式"鼓风机和移动式风扇(它是自动清洁式的)补充。这些系统的缺点仍旧是,飞花受干扰地分布在工作间。在封闭筒子架中或在针织机上收集与排除飞花的任务,主要是由 Memminger-IRO、Frohtich、Pilotelli 和 LTG 完成。在封闭筒子架中,主要以内部空气操纵的鼓风机系统是趋势。在针织机上有两种不同的竞争系统。Luwa 在平针织物机或罗纹大直径圆形针织机上方设有一种特殊空气出口,空气从该排气口以"帽"的形式向下流动,试图使纤维和飞花沉积在地面上。Memminger-IRO、Monarch 和 Shelton 选择在平针织物机上,用一个位于针织针筒上方的抽吸罩解决飞花问题。如果预测的飞花排除量可以获得,那么可以预期降低成本。

就针织机的质量监控而言，常规的技术并没有什么特别的进步，而是得到了升级或改进。为每台高速机器提供织物扫描仪成为惯例。

自1987年国际纺织机械展览会以来，圆形针织机的生产率已经到达了一个高水准。无需增加进给的数量（达到进给5次/英寸）或运转速度（在大直径机器上达到2 m/s）来提高性能。由于批量小，市场需求迅速，现在人们越来越重视高灵活性。复合针代表了一种技术进步。但额外的成本没通过改进舌针而得到补偿，它的性能已大大地提高。

第32课 经编及相关机械（一）

综合评价

经编机械包括以下分类：自动经编机、拉舍尔经编机、针织机、钩编织带机及一次性组件。技术的重点和改进的花纹组织，被少数市场领导者所支配。"专业"方面的供应商为利基市场服务。成本与销售价格的限制反应在技术标准中。技术可靠性与保持领先的发展是值得考虑的问题。

经编技术与织物工艺选择提供的结构和设计可能性，实际上是无限的。发展与创新表明这些优势被经编机械工业作为集中的挑战与机遇来理解。总的目标是提高竞争能力：

——适于布匹的最大织物生产速度，无任何设计要求幅宽多样，外观质量优秀。

——通过改变针织机构及其控制机构，通过改变喂给装置的类型与数目、包括衬纬以及在产品中加入别的纺织品（非织造网）与非纺织原料来开发结构与设计选择。

——创新拓展独特结构和设计效果适合于服装、家用与生活纺织品方面及专业技术用途的高质量纺织品结构。

——非破坏性的纱线转变

——在同一台机器中使用极不同类型和细度的纱线

——制造，需要三维产品、管状产品等，采用成卷技术。

这些技术和产品相关的属性是借助于工程开发（硬件）与扩展的软件系统实现的。

对于纺织工业其重点是在经济利益上，即高生产灵活性、高生产速度、质量保证与可重复性、快速设计变化、用户友好、环境适应性（低噪音）、电子花纹处理及自动化数据收集。

获得的成果是与纺织客户密切合作和最终用途的产品。

——纱线喂入系统

根据需要，纱线喂入系统以计算机控制的正向机械喂纱机构的形式存在，适合于高精度可重复的纱线喂入长度和高级的 EBC 系统，其提供连续的正向喂入形成花纹效果和控制喂纱长度。关于 EBC 系统，通过选择不同的丝线张力/线圈长度能获得辅助效果。新升级的典型 EBC 系统，是那些即使受总干线影响和发生电源故障，也具有更高的运转可靠性的伺服机构。

通过可编程伺服机构衬入纱线的垫纱系统的控制，提供高质量纬纱线喂入而纬纱上的应力很小。超过 5000 m/min 的速率不代表极限。某些解决办法，是充分利用喂入的纬纱长度来减少剪边的浪费，某些装置用于特别难织的纬纱（例如，玻璃纤维长丝纱）。

用衬垫纱线系统，可以用电子控制与调节装置来供给粗梳的带子和类似的纱线。与电子梳栉控制系统及织物卷取机构同步的这些喂入系统，在可再生质量方面满足最严格的要求，但必须是与纺织品兼容的。针编机在一定程度上也是特别的经编机，就喂纱系统而言，已经赶上了现代化技术。

——针织元件极其机构

从长远观点来看，朝向更高生产率发展的重点，仍然是在针织元件极其结构上。通过结合现代原料、轻量化结构、FEM 计算的梳栉轮廓和计算机优化处理进展，在这里获得了进一步的增加量（>3000 线圈横列/分，在高性能自动化经编机上）。

通过针尾弯曲段机构传动是所有的高速自动经编机的最新技术。值得注意的是，即在拉舍尔经编机（过去常常装备三角机构）中，为了低能耗、热量排放减少、低噪音、纱线应力小与织物质量好，需要有一个针尾弯曲段机构的转换器。

第 33 课　经编及相关机械（二）

经编机梳栉机构能由计算机控制的直线电机来驱动。结合 EBC 系统，花纹和样品生产的变化，垫纱循环次数实际上是无限的。用计算机控制，比用机械手段更能精确地给经编机梳栉分级。在驱动技术方面，未来的标准将由低噪音加连续可变驱动控制的带齿传动来建立。

为增加产品和形成花纹的灵活性及用户友好性,许多建议的解决办法提供了更简单的针织元件的调节(例如,从起绒或提花花纹向平纹织物的转变,将拉舍尔经编机上梳栉和栅状脱圈板的高度调整结合在一起)。

——生产数据收集系统

经编机的生产数据采集系统,总体上满足目前工艺水平。它们的目的是选择经济性(针织设备的效率与质量)和材料管理的最佳条件。它们的实用价值是没有疑义的。将电子控制扩展到整个经编机械群体及 CAD 系统开发的显著进步。常常复杂得多的经编结构(例如,由丝线张力创造的起孔效应)和适当地改变图案设计的技术控制要求,长期以来使得经编在此领域落后于机织和纬编。这个问题现在基本上已解决了。在监视器上逼真地模拟与打印输出自然色彩正在开发中。

——纺织织物的结构与设计

在经编织物中,修改织物结构和设计的选择,与其他技术一样多。在这方面,针编机和特种机特别值得一提,因为它们在家用纺织品与产业用纺织品领域中具有特殊的产品装置。

通过以下示例,展示了纺织工业在 设计的技术方面的明显进步,其产品结构与设计是前趋的,有时是完全创新的。

——有通过适合于快速反应的直线电机梳栉控制的自动经编机。对高端产业低成本生产具有重要意义,适合于褶间不同、短长度不同的弹性帘子布等。

——新的传动系统发展与提花机与多梳栉拉舍尔经编机

——通过压电提花部分(无通丝或磁力)控制单根丝线,适于平针或结构织物具有刺绣效果(针织锦缎无任何丝线的剪毛)。

——步进电机控制的提花梳栉

——以高生产速度、低噪音辐射和最大设计灵活性为目标

——在嵌花纬纱系统中,利用具有张力控制的几何级理想化的纱线喂入适于各种各样的产业用纺织品的品种开发(例如,双轴向半成品),织物卷取对拉舍尔经编机也能选择操作有机会获得感兴趣产品(中间衬料、家常服装织物等)。

——适于高速纬编生产线兼容纬纱加固的针编机

——5 框架弹力提花地毯针织机,采用无花纹绒线编织成花纹绒,外观清晰,耐磨性好。一项真正有价值的辅助选择为针刺与簇绒设计

——一种电子控制的拉舍尔经编机适用于非常厚重的产业结构,比如结网、纺织太阳能集成板、土工布。

用于监测不同机器功能的模拟显示值得特别一提。这是一个以多媒体技术为辅助的远程诊断的初始阶段。

总之，据预测由于其巨大的技术潜力，经编及相关机器在经济竞争中具有很好的发展机会。

除了已确定的最终用途之外，技术应用的扩展将越来越受到产品规格外形方面的影响。

高技术纱线的应用，在这里也正逐步变得更重要。

第34课　刺绣（一）

尽管席弗里与单或多头刺绣机之间的界限，就最终用途而言正变得逐渐不明显，但这两种系统在这里还将分开考虑。

席弗里刺绣机

在席弗里刺绣系统中，两种趋势是明显的。前者可居首位："万能刺绣机已经不适用了，取而代之的是专业化设备"。实际上，所有的席弗里刺绣机到目前为止，是以这样的方式被设计，即它们之间任何一种都能生产整批的刺绣产品，从透孔织物、网状物、纱罗和满地花图案、手帕、花边罩、窗帘饰边（流苏花边）和彩色提花窗帘布，到多色产品比如绒绣、边印及所有种类的徽章。对于良好的主机调定，只需要很少的手工调节来改变面线和底线的张力。

然而，目前的发展趋势被瞄向提供适合有限最终用途的机器，并装备这些机器以获得最佳的生产效率，否则是不可能的。这些之中最好的例子是 Lasser L77-115 和 Saurer "Epoca"。L77-115 具有 30.8 码相当于 28.15 米的刺绣长度，因而是迄今制造的最长的刺绣机器。由于采用了新的横向钩挂材料，27 米的长度正好符合用户购买窗帘的要求。

这些优点是明显的。没有交错的需要，机架简化为一次操作而不是三次（用10 码原料）或两次（用 15 码的原料）。但非常重要的是，连接是不必要的，因为尽管非常小心，在新刺绣与现存刺绣连接的地方会产生连接痕迹，疵点其很少能被补救，因此会导致降级。一个不可低估的因素是操作人员的技能水平明显偏低。

这种庞大的机器的缺陷是，它仅仅适合于 12/4 以上的重复，最大的速度比较适当的是 155 次/分，绣花架的高度只有 115 cm，以至彩色提花窗帘仅能必然刺绣到大约 110 cm 的高度不用重新调整。购买一台这样的机器，必然意味着获得了一种新的设备。

Part 2 Translations
参考译文

另一方面，Saurer"Epoca"建立不同的优先权。采用"Actifeed"，实现了一种全新设计的绣线喂入系统，刺绣速度达到400次/分。下一个针绣的精确长度由计算机控制，喂入罗拉(较早期的金刚包覆的罗拉部分)按合适的增量转动。规定与所涉及的材料相适应的正面旋转角度是可能的。与全新的绣线控制系统一道(仅有一个导纱钩)，这保证在刺绣形成过程产生在面线中的张力高峰值被大大降低，因此以最佳绣线张力运转，同时减少断线的发生总是可能的，尽管刺绣速度很快。另一个对此贡献者是全新的绣花架，它已被"无框架织物支持系统"所取代。

Xtraline 支撑系统适合罗拉双织物使用伺服电机从上到下有5点驱动和从侧面(正面在 Automaten 上)2点驱动。另一个新特点是，刺绣织物卷取不再依靠操作人员的触摸灵敏度，而是由"QuickRoll"系统控制。特别研制的传感器监测与保证原料的恒定张力。机器不再需要传统的是基础，换言之不必在适当的位置"铸造"。

这项新开发只是单层形式。它被在刺绣厂采用，大量色彩或原料能最有效地使用，使用达到良好优势的3040型而出名的 PentaCut 线头修剪装置，适合中到大批量的订货。时尚刺绣是一个重要的领域。相反地，对于单色相对粗糙的刺绣，比如流苏花边(窗帘滚边)和随后的单色染色及满地彩色提花窗帘是相当不适合的，也是极低效的。

电子筒子架控制

第二个趋势包括，普通的经过长期试验证明的标准机器的不断完善。对用电子筒子架控制精密的细节是理所当然的。除 Comerio 以外，循环与色彩变化对供应商来说现在是标配，线头修剪装置也一样。进一步的改进包括，因更大的框架高度和筒子架更宽的移动刺绣范围的扩大。另一个例子来自于激光，与非操作针类似，梭子也被停止，进而防止磨损和弄脏梭芯线(已不被需要)。

单与多头自动刺绣机

这里的趋势与席弗里刺绣机一样明显。一方面是轰动性的新开发，另一方面是对工业上成熟的机器的改进与完善。

第35课 刺绣(二)

带激光束的多头机器

Tajima 声称关于其对装备激光束的多头机器进行了引人注目的创新。激光束

177

被装设在每个刺绣头的侧面,借助于它设计的刺绣部分,比如文字其恰恰被绣在绣花边界线的外部能被修剪。例如先在白色底板上放绿色,随后放置红色材料。可以刺绣任何图形。该机自动地将原料横向移动到激光器的旁边并驱动激光器。它烧去了最上面的,换言之红色的原料距离轮廓线要求处。然后它向外移动一点儿并烧掉绿色原料,对红原料产生一种阴影效应。白色底板不受影响。

这些 TLFD 机器的最大优点是任何厚度原料能被裁剪,甚至在同一个设计中裁剪不同类型,没有边的松散。简单的"调色板刀"效果当然也是可能的。不过,激光照射的成本仍然非常地高,所以在可预见的未来,没有具有 20 头或更多头机器的可能性。预期有最多 8 头的机器,更为现实。

第二种在米兰引起轰动的是新开发 X25 240D 多头刺绣机,由 ZSK 制,有门架和边框及 25 头。专门为彩色提花窗帘的刺绣设计,具有大约 6 m 的绣花长度和最大 1000 mm 的刺绣区深度。图框配合需要的时间自然比在席弗里刺绣机花更少,但问题持续出在交错后的连接。Florentine 网状窗帘,具有连续图案的窗帘,因此是不合适的。但对这种机器感兴趣的那些人士来说,有一些非常明确的建议适用于在其他方面。

进一步的升级与完善

与席弗里刺绣机一样,在单和多头自动化方面经过长期良好试验与测试的机器得到了进一步升级与完善。一个例子是建立在绷子绣花机基上的多头自动化机器。用它可以高的速度(达到 600 次/分钟),针刺绣花链式线迹和桂花针法。它能被转换到链式针迹,因而创造了"凸纹"针绣,并能使狭带和类似物被缝上。一些机器制造商在双线系统上运转的每个标准头上加了一个绷圈刺绣头。

这些机器确实是通用的,但使用两种不同的机器是否更明显地适合针刺绣花器,将由工业经验来揭示。Pfaff 声称自己是"世界记录保持者",由于其机器是第一台也是唯一一台可以连续操作的机器,1000 针/分钟。几乎所有的制造商,都给他们的机器配备了依赖于针迹大小的电子速度控制系统,其中之一是"跳针"技术。因为非常长的针迹,刺绣框仅横向移动了一半,针没有在这个"暂时停止"中渗透。

这比过度地降低绣花速度更有效,否则刺绣速度对于机器上的速度和确保最佳效果是必要的。在这种情况下,"自由臂"模型不可被忽视。他们的目的,是仿效从美洲传到欧洲的时尚,绸缎边无舌尖顶帽的前部刺绣或人工短袖圆领衫和汗衫、手巾、全套浴室用品等。

要刺绣的物品被固定在特殊的框中。同时，一些机器制造商专门为这一目的装备他们的机器，另外是一些靠降低桌面适当地改装"标准"的多头自动化机器。最后需要提到一个趋势，即今天基本花纹图案与其他设计的刺绣正在日益变成印花与刺绣的结合。例如，印在短袖圆领衫上的图案是简单的，然后靠少数的不连续的部分"增强"。自然，这要求在精确地每个头上正确的位置记录刺绣的复杂技术。此外，还要有许多较小的公司提供常常是必须的辅助产品。这些中的例子是，Applicut 装置适合于"锯齿剪裁"掉在嵌花物品中的剩余的材料，光电读数的 Heinzle 高速循环机，无芯线轴和新颖提花地织物，它提供了大量的在胸袋刺绣中的覆盖针迹等。

第 36 课　编带机与梭结花边机（一）

横机在编带机领域，尽管有关机器是否实际需要，包括成本的争论仍在继续，但在安全、灰尘辐射和编带机噪音的降低方面都受国际标准 ISO11-11 很大影响，机械制造商的继续努力朝着自动化、高容量的筒子、新的原料及更强有力的过程监测来提高生产率。

来自 Cobra 的编带机设计为从非常细的编带到更小的绳索市场范围。Cobra250 型目前被制商要求以更高的速度运转，同时在采用低级质量纱的同时，有柔和作用的织花边机构的 Cobra450 能保持编带质量。新结构材料与 Cobra450-80 新花边筒子组合的结合，能使用非常细弱的纱线。比较宽的 80 mm 的锭距允许有细长丝纱的高筒子容量。Cobra 的花边筒子具有因更宽的筒管和更高的筒子容量，而筒管高度保持不变。目前，编带机的显著特点是密封与高标准的操作可靠性。

ETK Lesmo 编带机含有典型的安全性和降低噪音的特点。为了降低噪音及全面的安全性，包括退绕机构的所有运转部件被完全封闭。

Herzog 推出了范围从细的编织细绳到登山绳的生产机器。SE 1/16-180 新型编绳机，现在具有更高的 2208 cm^3 的筒管容量（以前为 1400 cm^3）。在机器运转的同时，一种新的驱动概念，为锭翼圆盘转动速度的电子控制、纱线退绕、和适于变换的用器具卷绕创造条件。被推出的最大的机器，也是为登山绳的生产而设计的，具有 336 mm 的锭距和用 16 个花边筒子装备，每具有 13271 cm^3 的筒管容量。捻距对整个筒子能保持不变，而这几乎排除了使用电子控制系统。LE 专利的编织眼显示，当芯丝已经用完，由于不规则的纱线张力或在编带中形成的结子退绕机构不正常工作而歪扭。一种光学传感器，在花边筒子变空时停止编织机。为监测生产效率,机器可配备生产信息记录系统。

Melitrex 开发的新双头编带机具有 1300 mm 锭距(锭翼圆盘速度 280～320 rpm)。筒子容量大约 800 cm³。该机装备有减噪装置。

1995 年研制的有 2400 mm 锭距和 16 个花边筒子的 O. M. A. 编带机,类似于新近开发的双头 104UCM/2 机器,目前配备了在纱线退绕电机与锭翼圆盘传动之间的电子自动同步装置。包括在机器启转时,这些能使编织常数与锭翼圆盘的转动得以调整。纱线退绕机构和锭翼能被以缓慢运动的方式分别前进和倒车,适合于纠正编织疵点。控制台显示锭翼圆盘的旋转速度、生产速度及编带卷绕角。在样机器的封装、成品率、铺设角和降噪效果良好,噪声放置达 75 db(A)。对于 240 节距的机器,一种新的花边筒子被研制,有 3200 cm³ 的筒子容量;为了维持更大的纱线补偿能力,一种在筒子的上方安装了一个带有多个绞盘罗拉的新型丝线张力系统。

第 37 课 编带机与梭结花边机(二)

新型的 Ratera 机器以修改设计的方式被推出。一种最新研制的编织眼监测器,如果退绕故障或因不均匀纱线张力歪斜时会关掉机器。适于这些机器有 80 mm 或 104 mm 锭距的花边筒子具有更高的筒子容量。耐纶嵌入生产商要求飞片盘中的尼龙衬垫锭子圆盘及两个花边筒子圆盘之间转动的套筒,要保证平滑的花边筒子运动的顺利传递,并降低噪声辐射,使花边筒子脚更坚固及锭翼圆盘减少磨损。

一台首次产品是三头编搓绳机,每个花边筒子接收 4 个筒子,生产装饰用编带,还有一个带有 20 个花边筒子的螺旋编带机。

Steeger 公司用其新设计的花边筒子托架轴承与梭结花边技术相似,导纱器及组件采用模块化形式,提供了解决问题的一种灵活方法,尤其在专业机器领域。由于轴承段的特别设计,编带机可以制造任何所需数量的锭翼圆盘,尤其对花边筒子数目超过 100 的技术领域的应用特别有利。

TrenzExport 目前还提供编带机的安全涂层。

Wardwell,是一家生产高速编带机尤其适合金属丝编织的制造商,推出了一台全封闭传统设计的机器,适用于更低噪音辐射和高安全性。

梭结花边机

Hacoba 开发了一种有电子退纱的梭结花边机。轴承与花边筒子的运动是建立在长期试验的原理基础上的。

  O.M.S.梭结花边机通过退绕罗拉有涂层的表面改善了纱线退绕条件。一种新型的电子控制的中央润滑系统，明显地改进了机器的维护。一项进一步革新是从编绕点移开的顺流的加热器元件，它在 80 ℃ 熔化易熔的胶合线。通过封装机器下部实现了噪声输出的显著减少。这个机器的新花边筒子设计非常简单，在花边筒子的外面没有易受磨损的导杆。

卷绕机

  近年来，值得注意的是，越来越多的编带机制造商参与到自动卷绕机的开发中来。因为人们普遍认识到，编带质量受花边筒子质量的影响很大。

  Hacoba 公司研制了一种新型符合人体工程学的四锭电子仪器控制自动卷绕机。Luigi(Hacoba 的意大利子公司)开发了一种新型自动卷绕机，能用按钮控制精确生产规定的筒子长度。借助于压紧方式，丝线头能被成功地起始。Herzog 提供一种电子仪器控制的四锭自动卷绕机，能自动改变筒子长度。

  卷绕机械制造商，也开发出了适于花边筒子的特殊卷绕机。Cezoma 公司推出的新设计的卷绕机有电子控制，该控制容纳多达 20 个程序并担负各种监控功能。为了控制黏接纱线长度，丝线张力被一个传感器臂(它控制下游丝线张力器)测定。

  O.M.R.公司开发的一台四锭全自动电子驱动，适合于生产平行卷绕和交叉卷绕。该机由一台直线电机来控制卷绕成形。而由 Ratera 公司推出的新型 BA-4 全自动 4 锭卷绕机，最长可达 200 mm。